THE CINDERELLA TREE

The Story of Mayr Bros. Logging

By Werner Mayr

Sandpoint, Idaho

COPYRIGHT © 1992 by Werner Mayr

Manufactured in the United States of America

ISBN 1-879628-01-5

All rights reserved by the author, including the right of reproduction in whole or in part in any form.

Published by:
Keokee Co. Publishing
P.O. Box 722
Sandpoint, ID 83864
Phone: (208) 263-3573

Edited by M.R. Compton, Jr.
Book design by Chris Bessler

First Edition

Dedicated to Jennie K. Mayr, my faithful wife and partner during good times and bad.

Table of Contents

Foreword
Page ix

Introduction
Page xi

Acknowledgments
Page xii

Chapter 1: Exodus
From Gossau to 'The Harbor'
Page 1

Chapter 2: The Twenties
Growing Up on the Wishkah
Page 11

Chapter 3: The Depression
Two Men and a Horse
Page 35

Chapter 4: The War and After
From Horses to Tractors, and On to High-lead
Page 59

Chapter 5: The Fifties
Cedar, Cedar, Cedar
Page 83

Chapter 6: The Sixties
Opportunity From the East
Page 109

Chapter 7: The Seventies
The Sky's the Limit
Page 135

Chapter 8: The Eighties
The Sky Caves In
Page 163

Chapter 9: Into the Nineties
Moving On
Page 177

Appendix
A Family Tree
Page 187

Foreword

The Cinderella Tree

OF THE FOUR MAJOR COMMERCIAL CONIFER TREES in the Grays Harbor country, the Western hemlock was the one overlooked. Douglas fir was of such density and moisture content that it could be used immediately as construction lumber, without pre-drying and held the number one spot.

The Sitka spruce and Western Red cedar each had their special properties. Cedar is used for roof shingles, fencing and timber under houses, where Douglas fir would decay rapidly. Sitka spruce is a very tough wood and was useful for spars, rowing oars, and wherever weight was a factor. In the early days of aircraft, planes were built of a special grade of Sitka spruce called "airplane spruce," knot free, fine grain and vertically sawn.

Hemlock was just in the way. No one wanted it in the early days. A logger working a hemlock "show" was ashamed of himself. Logger's and millmen's experience with hemlock was not very good. Logs dumped in the salt chuck floated so low, some of them went down if you stood on them. Others were "dead heads," one end in the bottom of the river and the light end sticking its head above water like a muskrat swimming up the river.

The early loggers used hemlock for building pole skid roads. Douglas fir was too big and too valuable. In one case, Carlson Logging Co. wanted to build a pole skid road across half a mile of swamp where there were no trees so they brought in several rafts of hemlock logs especially picked for skid road logs. After logging was complete, the skid road remained and in time rotted into the ground.

Often looking at stands of hemlock passed by years ago, I would see the trees from a distance, study the sad-looking, drooping branches and read "Nobody wants me." Each tree said it to me in a different way. No two were alike.

Hemlock is a hard wood, wears well and looks like oak flooring. High-quality, old-growth hemlock was used for flooring. Grays Harbor Commercial Company in Cosmopolis cut hemlock logs for use as box lumber during the 1920s. But, for the most part, hemlock was not desirable as a building material, and most hemlock stands were bypassed. It was an unwanted species.

With the coming of the paper manufacturing industry, hemlock logs were found to be ideal for pulp and paper manufacturing. Now, after all those years, the lumber mill people have learned how to mill, dry and surface hemlock, and hemlock has become a proud tree in the forest.

My friend Dan McGillicuddy coined the term "Cinderella tree," for our Western hemlock because what nobody wanted became the most sought after tree.

We came into the forest at a time when the Douglas fir was almost depleted, and some thought us crazy when we went after tracts of hemlock. The first tracts we logged were exclusively hemlock, and it was hemlock, the Cinderella tree, that built Mayr Brothers Logging.

Introduction

THIS WORK COVERS A LONG PERIOD OF TIME that begins in 1870. It is the story of a family that migrated to America early in the 20th century, and after a short while in Ohio, came west to the state of Washington. Dad knew where he wanted the family to settle, and that was in the wilds of Grays Harbor, just a few miles inland from the Pacific Ocean.

It was logical when we came here to work at what was at hand, and in Grays Harbor, with such a conifer forest as stood nowhere else in the world, that was the timber industry. Seven feet in diameter and 200 feet tall, huge Douglas fir, Sitka spruce, Western Red cedar, and our Cinderella tree, Western hemlock, stood in a dense forest from the West slope of the Cascade mountain range clear to the tidewater.

People settled in Grays Harbor to chop and saw out existence in this jungle of old-growth conifer saw timber. Dad worked his time in the forest and then Marzell, Uncle Frank and I put in many long years in the summer dust and winter mud.

Some people equate the timber industry with that of mining for minerals, coal and oil, and natural gas; resources that are not replenishable. When they are gone from an area, that is all, they're gone, but the soil which supports such fine old trees as grow here will support another crop of trees, which only need to be protected from fire and insects and other ravages of nature. The forest floor will regenerate with the forest.

The times from the great Depression through World War II saw many changes and we went through all of them, including labor union organization, beginning with the bloody organization in San Francisco of West Coast Longshoremen in 1934. Organization of the timber workers' unions began in 1935 and continued until most of the industry, including loggers, plywood people and sawmill people were organized and could for the first time speak with a voice that could be heard.

This welding together of working people in the timber industry brought them better working conditions and better pay, and only after 55 years is the union beginning to crumble around the edges.

Our industry went from cattle and horses through steam power in the 1930s and 1940s. Diesel power came in the 1950s. The trusty wooden spar tree was replaced with a steel spar rigged on a mobile carrier. No more whole tree trunks for spar trees, no more yarders on great sleds built out of huge 65-foot logs. The steel falling and bucking cross cut saws went the way of horses, cattle and steam, to a modern high-technology system. The old braided "jerk wire" whistle system was replaced first with a "tootsie bug," and now by radio-activated whistle.

The whole of life and industry has changed, and our industry is no exception. The old-growth Douglas fir is all but cut out. The new forest, much of which grew back by itself, is mostly hemlock. The droopy hemlock has taken its place in the lumber and plywood industry.

During the years after Jennie and I were married, I worked in the logging camps and for many years was away from home two or three evenings a week. Jennie always took care of our children and was always a faithful and devoted mother.

In all the years Marzell and I worked building up the company, we rarely worked on Sunday except during an emergency like forest fire or flood or a sudden bitter cold spell. I always said if we went broke in the end, at least we had Sundays with our families and the Sabbath to honor God on.

Acknowledgments

THANKS TO Mother and Dad.

We always received the loyal support of our parents. They knew our direction was our choice, though Mother could never understand how we could dare go into business for ourselves. In the old country in those days you followed your heritage, what your father did. But we were in America, a new land. Dad had been a logger and understood.

Help and inspiration from many people on Grays Harbor helped me to continue to write, including Frank M. Franciscovich and John Hughes, both of Aberdeen; Cousin Willie Gulde of Koblentz, Germany; Cousin Anton Losinger and his son Dr. Anton Losinger of Rohrbach Bie Freidberg, Germany; and Mr. Haapenen of Porter, Washington.

People kept after me, asking "When are you going to write your book?" and many others helped me with little jabs of encouragement. I sometimes found it difficult to concentrate and continue writing, almost as difficult as when Clarence Baker and I surveyed property lines on the Quinault reservation during cold, wet winter days but, like then, I would "dig in" and little by little, it went better. Many thanks to my many friends.

Heartfelt thanks for the faithful patience of my wife, Jennie.

Special words of appreciation go to my sister, Margaret House, who resides in Aberdeen Manor. Margaret's help to me was invaluable.

She remembered the early days of our family and gave me correct information on the Mayr family home in Gossau, Switzerland. Margaret is the one person who remembers Grandfather Sommer and the journey from Switzerland to Columbus, Ohio, and then on to Aberdeen one year later, when she was nine and ten years old.

THE CINDERELLA TREE

Chapter 1: Exodus

From Gossau to 'The Harbor'

WITH A FRIEND, MY FATHER CAME to America in 1905. He was from the village of Rohrbach, 60 kilometers northeast of Munich, Bavaria. His great-grandfather, Anton Mayr, born in 1817, was a "gutler," owner of a small farm, having milk cows and raising cattle feed and general crops.

His son, our great-grandfather, Alexander Mayr, was also a "gutler," in the area called Peuntenhausen.

There was born Grandfather Kajetan Mayr. His wife, Creszenzia, was born in an adjacent village. Kajetan was a carpenter and builder. Our cousin Anton sent us his hewing axe, found in the rafters of the barn on the farm the family still owns. My brother Marzell saw the place on his visit to Germany in 1987.

My father had two sisters, but he was Kajetan's only son.

Dad came to America to pioneer and make a fortune, while Mother was living in Switzerland with my sister Margaret. He and his friend, Mr. Focht, sailed out of Bremerhaven, Germany, to South Hampton, England. In South Hampton, he saw a placard reading "Don't go to Cosmopolis, Washington. Unfair to labor." That referred to Grays Harbor Commercial Co., of which Neil Cooney was the big Irish boss.

Dad and his friend sailed into New York and went through immigration at Ellis Island and were poured out into America. Dad was 28 years of age. In the following year, 1906, he worked in the wheat harvest in Kansas, and when the harvest was over, he wanted to see San Francisco.

He went to San Francisco by railroad and arrived in the fall after the great earthquake of April, 1906. He worked on the docks

1

Marzellinous "Max" Mayr (third from left) was the "wood buck" for this steam roader for Carlson Logging Co. around 1907

unloading lumber from the "windjammers" that came down from Grays Harbor with good old-growth fir lumber for use in rebuilding the city. San Francisco never forgot fine old-growth "yellow" fir from Grays Harbor and specifies the quality even today.

In the spring of 1907, Dad was again moved by wanderlust and headed north overland, going through Klamath Falls, he told us, riding a stage coach. Somewhere in the Upper Willamette Valley, he got on a train for Portland and then rode from Portland to Grays Harbor on the Northern Pacific passenger train. Dad, with his inquisitive mind, came to see the place that was "unfair to labor."

Theodore Roosevelt was President, and there were hard times on Grays Harbor, but there was always work if a person was gung-ho ambitious, which Dad certainly was.

One of his jobs was "wood buck" for a "roader," a big wide-drum steam yarder that spooled up to 5,000 feet of main line and skidded logs down a pole road to the salt chuck. That machine took a lot of steam which meant fuel for a hot fire under the boiler and water to make steam. In this job, Dad worked for Mr. Gus Carlson, as he did every time he worked in the woods in those early years.

Often Dad asked for a job in the cutting crew and always the answer was "no." They needed to keep that steam yarder with a full head of steam.

Once, when there was some kind of labor stoppage in the cutting crew, the foreman told him "Now you can go cutting." Dad said, "Hell no! If I can't work under regular, I will stay as a wood buck."

At the bottom of 1908, Dad had built up a stake and he headed back to the "old country" and for Gossau, Switzerland. Mother and he were married in 1909 in St. Gallen, Switzerland at a cloister named Maria Insedlen and Margaret had a dad.

Mother was tall and slender with dark brown hair and at least four inches taller than Dad. Dad was a short, stout, robust man with curly coal black hair, and a mustache which he always kept, never shaved off.

Mother's father, Werner Sommer, was a veteran of the Franco-Prussian war of 1870 and '71 and had a strong desire for drink. He married H. A. Baucher in October, 1872, and lived in the village of Eberhardzell in the barony of Wurtemberg. There were born to them Tante (Aunt) Hedwig, Tante Marie and mother Mathilda, the youngest. Grandfather Sommer was quite a large man, and his daughters were very handsome, with dark eyes and dark brown hair.

When mother was 10 years old, Grandmother Sommer died. The two older girls had already left home, working out in Mannheim.

Mother told me many times, how much her father drank with his war buddies. He lost his shoe repair business and had nothing left. Food was scarce and one time she was sure he brought home a skinned-out cat carcass for something to eat.

In the cold winter as a pre-teen girl, she went out into the forest, broke off dry branches and brought them home so she had something to heat and cook with. Grandpa would get sick from drinking and at her young age she had to take care of him. She loved him.

The two older girls, Tante Marie and Hedwig, finally got the civil authorities to look into the matter and the situation got a little better. Then, Grandfather Sommer took an interest in a very attractive lady and wanted to marry her. However, she was a strong woman so she said, "Sie Mussen zuerst das saufen aufgeben," which meant, "You must stop your drinking first," and he did.

In 1909, Dad opened a shoe repair shop in Gossau. Dad had served two years in the German army from 1898 to 1900 as a foot soldier, and after military duty, he learned the shoemaker trade. The family lived upstairs from the shop.

There was peace in Europe and prosperity. A year later was born little Mathilda, a true fair-haired German. In 1911 Margaret started school in Gossau. Her leather rucksack (shoulder satchel) for carrying books to school and back home was later with us for years at home on the Wishkah.

Gossau is not far from the Boden See, an inland lake bordering

Switzerland and Germany. People cross by ferry boat, back and forth. Mother, Dad, Margaret and the baby Mathilda would go over to Germany on Sundays to visit Grandfather Sommer and his new wife, or Grandfather Sommer would come over to Gossau to visit his daughter, son-in-law and grandchildren.

Grandfather also had two other grandchildren. Daughter Marie Gulde had a son Willie, who was at that time about 10 years of age, and little Ludwig was the son of his daughter Hedwig and Ludwig Boehme.

It was a lovely time in that beautiful part of Europe. Peace was in the world, but Dad was a well educated man and in 1913 he began to see what was about to occur in Europe. He made a decision in the summer of 1913 to leave Switzerland and Germany and immigrate to America, the new land he had explored years earlier. Mother had not the slightest idea what it would be like.

Dad and Mother sold everything that summer of 1913 and bought passage on the Hamburg-American line for New York. Mother was about four months along carrying me. They went by train to Bremerhaven, for passage on the S. S. Vaterland, largest ocean steamer at the time, with three boilers and apparently three screws.

The Vaterland was powered by steam-driven turbines and was designed and built as the largest passenger ship in the world. She provided the following accommodations: First Class - 752 persons; Second Class - 535; Third Class - 850 persons (that was our class); between decks 1772; and Besatsung, 1,234 (no berth, sit where you can). She carried a total of 5,143 passengers on that voyage over.

After World War I the vessel sailed for United States Lines, having the misfortune of being in U.S. port when America entered the war. Beginning in April, 1917, it was sailed under U.S. flag by the U.S. Navy. From 1922 through 1931, it was the largest ship in the world.

The ship sailed from Bremerhaven, Germany, in August for America. After Bremerhaven she probably sailed into South Hampton, England, and there accepted more passengers. South Hampton was a gathering point for many immigrants from Baltic countries and Scandinavian peoples.

Margaret remembers the voyage well. It took 15 days and all the while she and Mathilda, age 4, had run of the ship. To this day Margaret says how nice the stewards were. Even though their passage was not First Class, the stewards treated the children first class. They could have anything they wanted off the tables, which were always set up with fruit, candies, cakes and pastries.

The voyage was too soon over and up loomed the Statue of Liberty, as if in greeting, welcoming with open arms another shipload of immigrants. The ship arrived at the dock in New York and you might imagine 5,000 passengers hustling and bustling around getting their belongings together.

The Mayrs had several big trunks and mother's big wicker basket. Years later she said she had a dream before leaving Switzerland, and in the dream she saw this basket, full of clothes, being loaded into a "pig" on Carlson's skid road which was across the Wishkah River from where we moved in 1917. A "pig" was a hollowed-out cedar log that the roader pulled up the skid road by the haulback line. It was used to carry camp groceries and other supplies to the logging camp and cookhouse.

Coming to America was a shock for Mother, with eight-year-old Margaret, little Mathilda, four, and me on the way. She did not understand the new language. Dad did and that helped immensely, getting everything together and clearing Ellis Island.

Dad arranged train fare to Columbus, Ohio. That much money he had and some left over, with which he bought an Aberdeen ticket. They arrived in Columbus in September, 1913. Dad and Mother and the family parted company at the Ohio railroad station again for another year. The only way they would communicate was by parcel post mail.

Mother and the two girls had no place to stay, no money, and nothing to eat. As Margaret says 75 years later, "I don't know how Mother was able to do it, what with no understanding at all of the new language."

Mother found a room to live in and Margaret went to live in St. Vincent's boarding school. Mathilda went to St. Ann's orphanage to live and be cared for. Mother got work at St. Ann's so they all had a place to live and meals to eat.

Dad headed to the Northwest to find a job and a home for his family. He found work again in the Carlson Logging Co. camp somewhere on the Wishkah or East Hoquiam, bucking firewood for the big roader.

December 29, 1913, I was born in Mt. Carmel Hospital, at Columbus, Ohio. Shortly following, Mathilda came down with measles and developed pneumonia. The latter claimed her life. Mathilda was buried in a children's grave in Columbus, Ohio.

Mother suffered terribly with a new baby and losing four-year-old Mathilda and her husband 2,000 miles away. There were some good people there and they helped Mother. One whose name I know, Sister Hermania, could speak German.

In hardship and tragedy, life must go on. Spring came. Dad had a job and found a house, and in August he mailed train tickets to Mother in

Ohio and with help from the people at St. Ann's, she got everything together and boarded the train for the final trip to the Northwest.

The train headed west across the great prairie, over the Rocky Mountains and finally down the west slope of the Cascades to Aberdeen. Dad met the train with Mr. John Huffman, a one-time logger and farmer who came with horse and wagon to carry the family goods and possessions to the new house in this new place. Mrs. Huffman, Lillian, was the daughter of Civil War veteran Mr. Taylor. Those were good people. They helped with whatever was needed.

Dad rented a house just completed from Mr. Albert Stewart. It was about two and a half miles up the Wishkah Road from the end of "B" Street, which was also end of the streetcar line.

We were immigrant Germans and on top of that Catholics, but Dad spoke English and by the time we got to the Wishkah home, Margaret spoke and understood English well. That was a great help to Mother.

There was a country school a half mile up the road where Margaret was enrolled in the fall of 1914. The children came to the school from across the river and from up as far as the "long swamp." The swamp was a spot upstream where the road became impassable in the winter. The children above there went to another school. Ours was the greatest distance down-stream from the school. There were no houses below ours before the end of B Street.

It was a one-teacher, eight-grade school sitting on the side of the hill just above the river bank. In the early days, children came by rowboat. The men would take turns rowing them up or down, with or against the tide.

Many years later I did a story about that school. I interviewed Margaret's first teacher in the new land, Miss Ruth Locke, who lived up near Elma at the time of the interview. She had just two years normal school education, but George B. Miller, Superintendent of Schools at Aberdeen, said she was smart, willing and ambitious and could handle the job well.

The road was graveled by then and Mr. Sam Smith was contracted by the Aberdeen School District to transport the students to town. He did so in a little 18-passenger motor-driven school bus, carrying the children to school a distance of four miles. Mr. Smith was a kind man and on his own time would take the children with several parents up to a log pond on the east fork of the Wishkah for picnics.

While we were living in the green house, in 1915 and '16, loggers coming down from the camps on the stages would stop by and bring Mother some good things like ham or some fruit they had hijacked out

One teacher, eight grades: Margaret Mayr, second from left in first row standing (with the pigtails and the smile) was enrolled in the Wishkah school in 1914.

of the logging camp cookhouse.

Mother was terribly lonely in those early years on the Wishkah. In a highly developed country like Switzerland, where people had lived for centuries, houses, streets and roads had been in use for many years, but, on the Wishkah, 50 years before there had been nothing. There were no electric lights, so in winter on cloudy, rainy days it was pitch black. A person needed a hand lantern to go outside. In those first years I doubt that Mother ever went to town.

At high tide, water would come two feet over the land around the house and even over the county road. All traffic over the road would come to a complete stop until the tide went down.

In August, 1915, Marzell was born at home. Dr. Kinne, a Congressional Medal of Honor recipient from the Spanish-American War, attended Mother and delivered the new son. I was an infant in those days. Dad was working for Carlson Logging Co. three-quarters of a mile up the road, so he was able to come home evenings. Mrs. Hansen, a Swedish lady who lived on East Second Street in town, came out to help Mother for several weeks after the baby was born.

Dad had his eye on what had been the Hiram LaPorte place, now owned by George Graham, who was a sourdough in '98 up in Alaska. Graham had some money when he came down from Alaska. Years later Graham's son, Lewellyn, told how his dad came down, bought the place from LaPortes, and sent for his family from Nova Scotia. The LaPortes then moved down the river about half a mile to a place called Woods Eddy on the east bank of the Wishkah.

Mr. Graham was clearing land for pasture when he suffered a heart attack and died, and Widow Graham wanted to move to town. Her two children were grown up and had no desire to farm. Dad negotiated a lease agreement with Mrs. Graham and we moved to what became the home place for the family. It had good spring water and, to people from the "old country," lots of land. Work was close by and town was only one hour's walk away.

The place was unique, a piece of ground between the newly graveled road and the river at the Wishkah Boom Co. rafting grounds. It had three acres of high ground and the same area of tide land which always flooded over at high tide. The place had two barns, a chicken coop, and a root cellar. The house was a real old-timer, built about 1885 by Mr. LaPorte, who floated the lumber up from town with the tide.

I was still too young to remember the move to the new place in the fall of 1917, but Margaret remembers. We had a road on one side and river traffic on the other. The tugs used to nose into the river bank and drop off goodies for us kids. It seemed tugs always had lots of good stuff to eat and had a little kitchen in which they could cook steaks or make soups and whatever.

The house was situated on a knoll about 100 feet from the river and 400 feet from the road. All around the house was an orchard which must have been planted before 1890, with 30 different varieties of apples, pears, plums and prunes.

In those days it was open range country which meant people's stock ran at large. We only had to fence one side and that side had a Red cedar, hand-split picket fence between the place and the road. The other side was bounded by the Wishkah River.

My sister Margaret was 9 or 10 at the time of our parents' migration. When we arrived at our destination on the Wishkah River, Margaret already was able to speak and understand the English language. This helped Mother in those early years in America.

Immigrants were not in all cases accepted into the mainstream of life. The father, the wage earner of the family, was sometimes sought out for special attention and ridicule by "the blue bloods."

"Oh, those immigrants are flooding America" one blue blood said once too often, and Dad heard. Dad had had enough, so he said to Mr. Blue Blood, "When I came to America, I had a nice suit of clothes, a suitcase with a change of undergarments and money in my pocket. When you came to America you didn't have a stitch on your bottom." That settled the issue and they became lifelong friends, and even I was honored to know a nephew of this blue blood.

In February, 1917, a baby girl, Hedwig, was born and Mother again had Mrs. Hansen come up and help for several weeks and Doctor Kinne was attending physician at home birth.

America became involved in World War I in April, 1917. At that time Carlson Logging finished the logging across the river. There was a big shipyard at the end of B Street where wooden ships were being built out of green old-growth Douglas fir as part of the war effort. Dad worked there until the war ended.

I remember the shipyard vaguely. It had a board fence all along B Street and there was a tent city just up-river from the shipyard. I don't think it was there very long.

There were watchtowers, one on each end along the river, probably to watch for submarine sabotage. The ship knees were piled in a neat row at the very end of B Street ready for use in making a keel for another ship. At low tide today you can see all that's left, piling from the ways the ships slid to the water on.

The ships were built entirely of wood with two forward holds. From Grays Harbor, they were towed down to Portland for engine installation and fitting. They were propelled by a single screw powered by a coal-fired steam engine which sat in the stern section. Many similar boats were later used for coastal trading, carrying lumber from Grays Harbor to San Francisco and southern California. By 1936 they were mostly all gone. Steel-hulled vessels with the engine amidships took over the shipping lanes.

In 1919, Dad went to work at his trade as a shoemaker, working for

Jeff Garman, a very early Aberdeen pioneer and businessman. In 1904, Mr. Garman had built a new two-story sandstone building on Market and G Street. At first it housed the U.S. Post Office and Mr. Clark's Grays Harbor Post. Also, upstairs was a German candy maker.

Being a trained old-country shoemaker, Dad could make a complete pair of dress shoes from leather; uppers, tongue, and foundation. He built shoes for some of the business men in Aberdeen before manufactured shoes came into vogue. After that, he repaired shoes, putting on half-soles and heels and screwing caulks into the bottoms of logging boots. This is what he did through the '20s.

Chapter 2: The Twenties

Growing Up on the Wishkah

IN 1920, DAD AND MOTHER established a home for the family on the George Graham place, two-and-a-half miles north of Aberdeen on the Wishkah road. It was homesteaded by Hiram LaPorte in 1877, who came up the river by dugout and put ashore on a bit of high ground where there was fresh water available and a safe place to build a dwelling and barn for stock.

The house LaPortes built faced the river. It was of box type construction set on cedar blocks. The rough boards inside were covered with cheese cloth and wall paper. This first section was 20 by 30 feet with an upstairs and a little unfinished hallway upstairs. The Grahams later added a kitchen and pantry with a woodshed, 16 by 20 feet.

All houses built along tidal streams of Grays and Willapa Harbor faced the river. The river was the road. In the years we were children there was always some kind of boat going up or down the river.

Many years before I can remember there was Captain Henry Spoon with his freighter, the "Aberdeen." Captain Spoon was a good friend of Mom and Dad and had his eye out for us youngsters.

Captain Spoon lived at the corner of 6th and Broadway in a circa 1900, two-story house with drop siding. It was later torn down and a mansion was built at that location by his son Donald.

Mr. Spoon came to Aberdeen with his brother from Michigan in the 1880s to build a lath mill for West and Slade on the north bank of the Chehalis River just above the mouth of the Wishkah. Soon after, he began freighting on the Harbor from Cosmopolis to Aberdeen.

In the early years, before the 1920s, Captain Spoon went up the

The family on the Wishkah. Mathilda Mayr holds daughter Hedwig. Next are Marzell and Werner. Margaret holds Werner, and Max stands by.

Wishkah river clear up to the forks and delivered supplies to the early-day logging camps and ranchers along the river and some to our place. From early 1900, he ran from Aberdeen up the Wishkah to Turner's Landing.

The Turners had a wayhouse where loggers and ranchers waited for a boat ride to town. The Wishkah Road was not graveled to accept winter traffic until 1916. Roads with no gravel and trails became impassable in rainy weather, and water transportation was dependable.

Captain Spoon had several tugs. One was "The Flyer," built by Swen Johnson, who lived on the east fork of the Wishkah. It had twin gasoline engines. After a short try at running freight, Swen sold the tug to Captain Spoon.

"The Aberdeen" was gasoline powered, and was renamed "The Clam." With it, Captain Spoon nudged the grocery scows up the river to the Donovan Corkery Logging Co. rail terminal.

Captain Spoon always planned to make it to the gap where the tug would enter the open channel on the flood tide. From there on up, it was a distance of about two miles to the terminal, where the scows he pushed could settle with the tide onto a gridiron, a set of piling with caps on them that were covered at high tide.

He had to thread his little tug and tow between the big fir logs, and if there was a fish net caught in the propeller or if he got hung up and was late in getting to the grid at rail terminus, he was really in trouble ducking among the logs. He could not leave the scow loaded with supplies just anywhere in the shallow river, which was full of dead heads and sunken logs. They would punch a hole in the bottom of the scow. Once a week he also brought up an oil scow, with oil used to fire the logging locomotives and later the big 13 by 18 steam yarders.

Most of the time Captain Spoon was by himself, pilot, deckhand, and engineer. I can still see him running back and forth on the little tug to change the bite of the hawser. Wherever the channel was wide enough, he would push the scow "side saddle." It gave him better control of his tow.

Our home was at the end of the open channel on the river and from there up Captain Spoon would tighten up his bite and slip between the outer log boom race and the left shore. Coming up on the flood he knew he had more water coming to float him over obstructions. If some stump rancher left their row boat hanging out in the river, Captain Spoon had to be careful not to smash it.

Very early, until about 1926, there were two steam-powered tugs towing log rafts down the muddy old Wishkah, the "Forester" and the "Ranger." They both had full captain's cabins for better vision.

A steam engine doesn't make much noise. The churn of the propeller makes more noise than the engine, and both tugs could almost slink up the river. In front of our house was a reach a little over half a mile long, and as children we knew about when to expect a tug. They came with the tide, and every day, the tide came an hour later. The tugboat activity going by our house from the earliest we can remember had an impact on us. It was like it must be to live along a railroad and see trains go by.

It was exciting and we would wait to see the action. We finally would see the "Forester" sneaking into view around the bend. Pretty soon, maybe halfway up the reach, the boiler on the tug would pop off with a roar, scaring the hell out of us, sending a plume of steam across the river. If the pop-off valve let go near our house, I don't know what we would have done, our house being only about 100 feet from the river channel.

In the first years, the "Forester's" boiler was wood fired and the wood was stacked around the pilot house. Coming downriver, with three rafts of logs hooked one after another, the tug pilot wanted all the flood he could get to get around the bends. He wanted to get as far as he could

The Forester in working clothes.   Photo © Jones Photo Co.

downstream before the tide started to turn, so they needed all the steam they could get. I can still see the slight wiry engineer and fireman running back and forth, face, hands and clothes covered with engine grease.

In subsequent years, different tugs came up after logs. These were smaller, gasoline-powered tugs; the "Flora Brown," the "Hustler," the "Bear," and Cap Mercish's "Tillicum," a World War I navy tug with an iron hull. The "Tillicum" spent more time on the river bottoms of Grays Harbor than the wood-hulled tugs.

Most people thought, as we did, this log movement activity would last forever, but the heyday lasted just ten years.

Dad bought the shoe repair shop from Mr. Garman about 1922. As part payment for rent, he agreed to keep up the wood furnace which was used to heat the whole building.

A Mr. Zeigler also was working as shoemaker, but he soon retired. I don't remember him but Margaret does. Sometimes he invited her to go down to Westport on a cruise aboard the "Harbor Belle" or the "Harbor Queen."

Dad's shoe shop was not over 16 feet wide by about 30 feet long, and had a shoe buffing machine against the back wall and a treadle machine for sewing uppers. He could sew heavy leather soles on the electric machine.

There was a narrow stairway going up to a loft where Mr. Garman

The Hustler on the Harbor, decked out for the Fourth of July.
Photo © Jones Photo Co.

had a little office. He must have been a fairly well-to-do person at one time. He was, I think, already in his sixties in 1920.

There was a row of four chairs for customers waiting while repairs to their shoes were made. Dad would keep his machines humming away and the guests would visit among themselves. In those chairs sat many of the old timers. I remember Mr. Graham, who came to the Harbor with the Wests from Michigan. Sam Benn, on his way downtown, would stop in for a chat, and loggers and ranchers would stop for a visit when they were in town.

They would sit there and log and farm and pioneer to beat hell. Sometimes someone would even come to get their shoes fixed. Dad had a little leather settee for them, so as not to displace the men in the chairs.

People came in from time to time with great money making schemes, possibly to hoodwink the foreigner shoe maker. Once, a gentleman came in really primed with a hot money-making idea. He just needed start-up capital and offered Dad a great opportunity. Dad took off his shoemaker coat, slipped on his jacket, put on his hat, and said, "Follow me."

They started down the sidewalk, the man still giving forth great superlatives about his plan.

After a block or so he asked Dad, "Where are we going?"

Dad answered, "If your idea is as good as you say, the bank has money to loan. They can help you."

The man sped up and disappeared down the sidewalk.

The home we lived in was about two and a half miles from the shoe shop and Dad walked both ways. For a time he rode with Harry Gibson, our neighbor, who was business agent for the Carpenter And Joiners Union.

Our house had running water inside but no inside toilet or bathtub, only a kitchen sink. The kitchen stove was a "Royal Oak" wood-burner with coils in the fire box for making hot water. A galvanized hot water tank of about 15 gallons sat alongside the stove.

In the main room, in the older part of the house, was a wood heater that was allowed to "burn down" at night. Fire was feared much, and both chimneys were bracket chimneys with no flue liner inside. Some mornings in winter it was so cold in the house water froze sitting in a glass jar on the windowsill in Marzell's and my bedroom, which faced the river. There were no electric lights until 1926. We used kerosene lamps and lantern for outside for years and finally Coleman gasoline lamp and lantern.

The house did have a telephone. It was hooked to the Wishkah Boom Co. line. Our number was 4F4 and we answered four long rings. There were about six phones on the four-wire line. Another four-wire line belonged to Grays Harbor Logging Co., which was splashing logs on the east fork of the Wishkah.

My first year in school was at St. Rose Academy, which sat where Aberdeen Manor does now, on North G Street near the old St. Joseph Hospital. Margaret was in the seventh grade there. The teaching Sisters were French-Canadian, and being children of German immigrants made for some rough times.

Sam Smith will be remembered by those who came in contact with him when he was custodian at the lower Wishkah school. In addition to bringing children to school and tending the furnace, he always brought a big jug of hot cocoa for youngsters who didn't get any breakfast. When the lower Wishkah school closed, Mr. Smith transported children to Aberdeen schools with his bus. He did not discriminate against children going to a school of their parents' choice. He owned the bus and nobody was about to tell him what to do.

People who do things like that will be remembered at the final roll call, and those sons-of-bitches that came later and said you couldn't ride the bus because you went to a Catholic, or Jewish, or Seventh Day Adventist school and discriminated against children will roast in a special place prepared for them.

Before the public school system provided direct transportation for families on the lower Wishkah into Aberdeen, there were others who furnished transportation, including Andrew Johnson and, later, his son-in-law, Armus Tikka. When Armus Tikka took over in 1921 or '22, he bought a new long-frame, one-and-a-half-ton Ford chassis and built the bus cab out of wood and canvas, a far cry from today's buses.

Our parents wanted us to get a good education. St. Rose closed in about 1924, and we began going to public school. In 1927, St. Mary's opened a brand new school on North H Street staffed by the Dominican Sisters. Our parish priest, Father Michael O'Donnell, came by Dad's shop and asked if Dad would bring his children back for a Catholic education. Dad and Mother agreed. Hedwig and Marzell went back and both graduated from the eighth grade at St. Mary's. Margaret went into nurse's training at St. Joseph Hospital and received her State Nurses Certificate in 1928.

I was in the seventh grade at Terrace Heights and stayed to finish out the year and went to junior high for eighth and ninth grade.

When Dad got his 1921 Dodge touring car from Mr. Wiser, we became more mobile and had a lot more flexibility going to school. I remember well how cold it was riding in that car to church on Sundays on cold, rainy days. The car had wooden spoke wheels and narrow, high pressure tires.

Once in a while, Dad would take us on a Sunday drive to see some of the folks' German friends. One such jaunt was up the East Hoquiam River and around through Hoquiam. I remember going along the east side of the river by the E.K. Wood Lumber Co. mill and seeing two windjammers tied up at the mill docks. That must have been around 1925.

Another time we went toward Montesano on the old highway, where the car could go 30 miles per hour at top speed on the straight stretches. The highway was paved, but the Wishkah road was gravel.

When working at his shoe shop, Dad always parked around the corner on Market Street. It was laid out for angle parking during the 1920s. Market Street one time had a park strip in middle of the street between G and I streets, with a couple of World War I cannons on display with flower beds in between.

Across the county road from the house we lived in was the remnants of an old-growth hemlock forest with many huge old trees scattered throughout. It extended clear over the top of the hill. It was a great place for us to peel cascara bark and hunt for flowers during spring and early summer.

In spring, we heard the blue grouse male hooter and along the creek bottom would be the ruffed grouse drummer creating their mating calls. I don't believe there was anywhere in the world a more peaceful or happy place for children to grow up.

The Wishkah River always was a source of intrigue for us. We wished we had a row boat. Finally, by some trade, at about 10 years of age we got an old row boat and a couple of paddles, so we could float up with the tide and sit in back of the boat looking down in the water, our nose about two inches from the water, watching the fish swim by. Floating on the river beat walking up the road.

Marzell never seemed interested, but I was completely fascinated by fishing for sea-run cutthroat in the fall. I fished right off the river bank behind the house, usually on the first half of the incoming tide, and sometimes I would catch six or eight nice fish, 12 inches long, fresh from the ocean.

On the days we went to town during the day, usually walking, it was always more fun to get back home again. There we were free; free, like a bird.

Going to Aberdeen on July 4 was something to remember. In the 1920s, there were fifteen railroad logging camps putting fresh-cut logs into the water of Grays Harbor.

When these camps shut down for the "July 4 Splash," there were a thousand men standing on the north sidewalk from the Heron Street bridge up to K Street, logging to beat hell, telling each other all about the haywire, gunnysack outfit they were working for. It was the same as now. Equipment was always broke down, the saw filer was no good and the camp cook was a gut robber.

A few of the companies were Donovan Corkery (Hardtack Bill Corkery), Saginaw (Dirty Old Saginaw), Polson Logging (I worked for Papa Polson), and Cedar Swamp Mr. Smith, of M. R. Smith Co. at Pacific Beach.

True to form, loggers did their best logging on the sidewalks of Aberdeen and excelled in their sexual exploits when at work in the logging camps.

During the years between 1920 and 1929, Aberdeen was synonymous with loggers, logging, logs, saw mills, and fresh cut Douglas fir lumber. How many were working at the industry in those days? I venture a guess of 6,000. More than 80 percent of the loggers were single men, largely immigrants from the old country. They left Europe when they were 17 to 19 years old, between 1905 and 1915. Dad came over in 1905.

Loading Douglas fir at a railroad logging "show."

In the 1920s, Grays Harbor had two industries; lumbering and moonshining. Stump ranchers living in the valleys up north of town worked a little at a logging or mill job, did a little farming, milked a few cows, and made and sold moonshine or "rot gut." I remember several loggers died from improperly fermented mash or alcohol and formaldehyde "fusel oil."

Carl Foshaug, a booze trader, was blind at 41 years from "poor man's booze," and took his own life. Henry Lagger, a saw filer, died from the same medicine, as did others. The boommen picked up a cadaver just above our house once who probably died from the same cause and was hauled out of town for the convenience of someone.

During those times, a very famous tugboat operator would take a tug out over the "bar," (not the booze bar, but the Grays Harbor bar) and some miles out he would meet a vessel from somewhere north that happened to have aboard a goodly cargo of name brand Scotch, bourbon, etc.

The purpose of this meeting was to transfer from the foreign vessel to the tug all of the bottled contents. The tug carefully, so as not to damage any of its precious cargo, sailed back into Grays Harbor and, in dark of night, transferred the cargo to the basement of a famous Aberdeen church where it would be safe from marauding "feds." This was for the elite, so that lawyers, doctors, attorneys, judges, bankers, ministers, and the like would have access to the good stuff.

Our home place was surrounded by grass fields and an orchard of fruit trees that have since died back or fallen over. Marzell still has a few Italian prune trees which propagate from the root spur. They are not a grafted species.

The old place had an excellent water supply. That water was the best. It came out of sandstone rock at the base of the main hill. The system was put in as a joint venture by the school and Graham. The water came from a year-round stream across the road from the place. The water was captured in a little flume made of one-by-six lumber nailed together to form a V-shaped trough which ran to a 1,000 gallon wood tank. The tank fed a one-and-a-quarter inch galvanized pipe and water flowed by gravity through the pipe to our house and to the school house, where now living was the Coolidge family. They had three children.

In the winter, during or after a heavy rain storm, the water supply would plug up and stop and at ten to eleven years old we would be required to go up to the water tank to solve the problem. It was about a half mile from home and usually it was pitch dark and we went up through the woods to the tank carrying a kerosene lantern. There, we would pull the plug at the bottom of the tank, clean all the dirt and debris out of the flume, then climb in the tank and, with an old broom, sweep out and clean the tank. Then we drove the plug back in and watch for a while as the tank filled again with clear, cold water before heading back for home. Mother would have something warm for us to drink and a piece of German Afelstruddle.

The place also had a root house like most people in the country had in the early days. It was dug into the side of the hill, and in it were stored potatoes and other root vegetables, apples and pears, and one- and two-quart jars of canned goods on shelves.

There was no refrigeration and no ice box. Fresh milk was kept cool on the shady back porch in a cabinet with screen on the sides and the door. This is where milk would sit in pans so the cream could be skimmed off.

Most people living today don't know the basics of having a family milk cow. To keep us in fresh milk and cream, Dad milked the family cow morning and evening. It was Marzell's and my job to bring the cow in for the evening milking. She was kept in the barn all night.

A milk cow needs to be impregnated every year, usually three months after she has calved. After the required time, she has another calf and is "freshened up" and gives lots of milk. That was the days before artificial insemination, and it was our job to walk the cow up the road to either Tikka place or Teitge place to see her bull friend.

When we were small, between four and seven or eight years, we were warned to stay away from the river or the evil man who lived under the water in the river would come. He had a long pole with a hook on the end and he would pull you under. That worked pretty good until we got a little older and got closer to the water.

The neighbor children, the Coolidges, splashed in the river below our place on the old gravel bunker area. Gravel was spilled there and we didn't sink up to our knees in the mud. Finally we learned to swim or, at least, keep above the water. Later we swam at another spot with neighbor kids Gene and Lynn Short.

By 1925 we had begun berry picking for Mrs. Horton. She had a raspberry garden and we would go and help with her berry harvest. The following year we began picking logan berries for Andrew Johnson and later Sadie Tikka.

The year 1925 was dry and brought forest fires. A particularly bad one was in the Wynooche-Satsop area of the Schafer Bros. Logging Co. operations. The fire spread and ran into Wynooche Timber Co. operations and lasted several days. It darkened the sky and dropped cinders out of the sky.

Such fires were not unusual during the 1920s with all the logging going on in the county. Many of the huge steam yarders were still wood-fired and the exhaust went through a smoke stack. On a long hard pull, the burning cinders would fly out of the smoke stack and spread around the yarder and loader. In those days, hardly anybody shut down for dry woods conditions. Rather, they kept on logging until the fire lapped at their heels.

Around where we lived a few small loggers still took out patches of hemlock with a sort of skid road. The cables, turn and everything, just went straight across the road. The cars and other vehicles had to stop and wait for the cable to slack off. This type of logging occurred in four places close to where we lived between 1922 and '24.

One logger named Jake Vohs logged timber off the Horton place with a sky line. He logged about 12 acres of old-growth hemlock and one old-growth fir and tight-lined them right over the county road. I suppose they had a flag person there. Russ Ellison told me his step-dad hoped the sky line would break when they swung the big fir butt off the hill so he could get some handy firewood. Sure enough, the line broke and dumped the log in Horton's back pasture for the winter's wood.

In 1925 the logging industry was going through many changes. By then, most logging was to railroads. Ground-lead logging and the big road yarders were no more, and spar trees and the huge steam yarders were coming in.

This was high-lead logging with steam.

A system was devised using a wooden spar tree as tall as possible, up to 180 feet, with a set of blocks (pulleys) at the top. Through these pulleys ran two cables, the "main," which was one-and-three-eighths inches, and the "haulback," a "mere" seven-eighths of an inch. The spar tree was guyed by seven top guys and five buckle guys of the same size cable as the main line.

Downhill from the spar tree, several sets of blocks were attached to stumps or trees, and by means of a small cable called a straw line, the haulback was threaded through the lower blocks and attached by shackle to an eye on the working end of the main cable.

At that end of the main was the "jewelry," two or three "bells" by which 28-foot chokers made of one-and-an-eighth-inch cable could be attached. The chokers were the means by which the "turns," or groups of logs, were attached to be dragged up the hill.

Powering this affair was a steam yarder on a wooden sled, tied down to two of the biggest logs available with one-and-an-eighth-inch cable. The yarders were wood-fired with an extended fire box, and later ran on crude oil when rail lines ran close enough. They had two steam cylinders as large as 11 inches in diameter with an 18-inch throw driving a shaft with a gear on each end that, in turn, transferred power through a cylindrical clutch system to the winch drums. The yarder had three drums, or cable spools; one for the main line, one for the haulback, and one for the strawline.

To raise and lower the cable, the engineer would "slam on the friction," then regulate speed with a throttle and brakes on the drums.

The main cable was used to bring the "turn" to the loading area. The haulback cable was used to "haul back" the main cable to the logging area.

In charge of the rigging crew was the hook tender. His assistant was the rigging slinger, and under him were three or four choker men, the engineers, the chaser and the whistle punk. The hook tender, rigging slinger, and choker men worked down-hill from the yarder. The engineers ran the yarder and loader. The whistle-punk, watching and listening for the downhill crew, used a braided jerkwire to pull a steam whistle that signaled for the engineer. The chaser worked unhooking the turns at the decking area.

With a full head of steam, the weight of the huge diameter main line and large drums of these yarders gave an awesome pull when a big log was hung up behind a log or stump.

When this would happen, the hook tender, would send in a ho-ho whistle, which told the engineer to pull the rigging and sometimes the turn back a little; then he'd send in a ho ... Ho!, which meant hold a tight haul-back. The engineer would jam on the main line friction, lift up the steam throttle and when that cable got tight, one of three things could happen. Usually the turn jumped the hang-up. Maybe a choker would break, or, if the main line was worn or had a bad spot, it might part. Rarely, but sometimes, things would come to a complete stop and a there would be a calm, and we would think, "Now what the hell do we do?"

In any of those events, you would see a fully-rigged Douglas fir spar rigged to a steam yarder, main line tight, the spar tree and rigging just shuddering. It was a sight to behold.

We were about ten years of age when the last splash dam blew out. It was on Big Creek and belonged to Aberdeen Logging Co. They were logging fire-killed snags. It washed logs and parts of the dam clear down to where we lived.

The Malinoski dam on the main fork did not suffer that ignominious fate. Like a stately patriarch, it held all its splashes and was scene of some happy camping trips in the late 1920s. There was a hollowed out cedar log dugout in the pond we paddled around with while we were on a camping trip during 1928. It held water until 1929.

At the age of 13 or 14, while Marzell worked at home in our garden and berry patch, I began to work for Mr. John Huffman in his truck garden. Mr. Huffman was a Missourian, and he and his wife, Lillian, had three grown children.

How he coaxed garden vegetables to grow out of that wet, sticky soil was near a miracle, and all the love I have for gardening I learned from Mr. Huffman. I learned to plow with a horse, till with a harrow, and disc plant with a hand-pushed wheel seeder. He grew fresh vegetables like onions, radishes, lettuce, carrots, beets, and cucumbers, plus a lot of other good stuff out of that wet soil.

At the time, Aberdeen was booming, and Mr. Huffman had a one-ton Model-T Ford truck. All summer and fall he would head for town at 6 a.m., his truck loaded with fresh vegetables. He was a truck farmer. He was not much for growing berries.

Did you ever weed and thin long, seemingly endless, rows of carrots, radishes and beets? I have been there. I made $1.05 a day and I don't remember how long the hours were, but I remember Mr. Huffman would sit in his rocking chair on summer afternoons where he could see most of his garden and whether I was goofing off, busy working or had sneaked off. Sometimes he hired extra help, but mostly that kind of help was just not available. Most youngsters my age didn't want to or didn't have to work.

Mr. Huffman had a unique irrigation system using water which came up the slough from the river at high tide and under the county road to Huffman's field through a drainage ditch which ran alongside the garden. There, a centrifugal pump driven by a 7 h.p. electric motor pumped water through a two-inch pipe on which were mounted sprinklers.

I looked after the pump at night, because that is when tides are high in the summer months. When the tide ran out, I shut the pump down and went home.

After the October, 1929, stock market crash, mills were closing and logging camps as well. The good market for Mr. Huffman's fine, fresh vegetables suffered along with the rest of the economy, and Mr. Huffman kind of gave up by 1930.

Next to Huffman's, a logger named Big John Johnson ran a place where loggers could get booze and women. In the summer, when logging camps were busy, around 3 in the afternoon I would see Johnson's 1927 Oldsmobile coming up the narrow lane which lead into Huffman's barn area and Johnson's house. The car would be loaded with thirsty loggers.

After a while, Big John would head for town again and maybe take some of the spent loggers back. Loggers came and went from Big John's place from late afternoon until midnight lots of days, especially Friday and Saturday.

In 1925 our sister Margaret lived with the Sisters at St. Joseph Hospital while she was in nurse's training. The hospital was in the new red brick building built in 1919 and '20.

Margaret joined the Dominican Order of Sisters in 1927. I rode with our parish priest, Father Michael O'Donnel, to Everett for the celebration of taking the vows of a novice. Several other young women also took their vows and then wore the harsh, strict, mostly black habit with a starched white collar.

After Margaret joined the Dominican order the only time we saw her for years would be during visits with her in the convent quarters at St. Joseph Hospital. These quarters were in the old, wooden part of the hospital. Margaret received her state-registered nurse's certificate in 1928.

Margaret was still not a U.S. citizen, and though she was born in Switzerland, she was not Swiss. Being of German parentage, she was a

German by origin, and the citizenship process took years. Had Mom and Dad received their U.S. citizenship before Margaret was eighteen, she would have automatically become a U.S. citizen.

Margaret told me she got no help from the head sister at St. Joseph. They were apparently still hung up on that World War I German hatred which was much worse after World War I than after World War II.

Finally, after years of writing and trying to get people to understand that the treaty between Germany and Switzerland was different than that between the U.S. and Canada, (If your parents are Canadian and live for a time in the U.S., a child born to the couple is a U.S. citizen.) the Superior Court Judge in Montesano awarded Margaret U.S. citizenship in 1935.

After a few years in the convent, Margaret, with another Sister, was sometimes allowed to visit her parents, brothers and sister on a Sunday afternoon. We used to take them for a rowboat ride on the Wishkah River on nice summer afternoons. These are moments of the growing up years that made lasting impressions.

Our Christmases at home all through the 1920s followed the old German traditions. We had a little mini-celebration around the middle of December in honor of St. Nicholas. On the evening of that day we would find cached around the house goodies like cookies, nuts, and fruit. In the old country tradition, if children were good, St. Nicholas would reward them. That may be where and how the Santa Claus tradition started.

The celebration of the birth of the Christ child was the big day. Mother prepared for that day weeks ahead by baking such goodies as Anise Kuchen, Gugglehopf, Leib Kuchen. The house, the yard, and all around were spruced up for Christmas. Dad, while we were too small to get a Christmas tree, would bring in a Sitka spruce, the tree with the sharp needles. It probably kept us from fooling around with the candles and Christmas decorations. We did not have electric lights so we used wax candles set in little candle holders that snapped on the tree limb and held the candle firmly in an upright position.

After all the chores were done and kitchen cleared up, we would all gather around the tree. Mother or Dad lit the candles, making the Christmas tree and its ornaments fairly glitter, a sight to behold. Everyone had an eye on different parts of the tree to be sure no candles burned down and created danger of fire. Mother would lead us in German songs like Stille Nacht, O Tannenbaum, or Edelweiss,* and others which I have forgotten. Before we had the Besharung, the opening of Christmas presents, all the candles had to be put out and the kerosene lamp turned up.

* Translations of the songs (from previous page)

Stille Nacht, Heilige Nacht (Silent Night, Holy Night)
Silent night, holy night.
All is calm, all is bright.
Round yon virgin,
Mother and Child.
Holy Infant so tender and mild.
Sleep in heavenly peace,
Sleep in heavenly peace.

Silent night, holy night.
Shepherds quake at the sight;
Glories stream from heaven afar,
Heavenly hosts sing alleluia,
Christ, the Savior, is born!
Christ, the Savior, is born!

Silent night, holy night.
Son of God, love's pure light.
Radiant beams from Thy holy face,
With the dawn of redeeming grace,
Jesus, Lord, at Thy birth,
Jesus, Lord, at Thy birth.

O Tannenbaum (O Christmas Tree)
O Christmas tree, O Christmas tree; With faithful leaves unchanging.
Not only green in summer's heat; But also winter's snow and sleet,
O Christmas tree, O Christmas tree, With faithful leaves unchanging.

O Christmas tree, O Christmas tree; Of all the trees most lovely;
Each year, you bring to me delight; Gleaming in the Christmas night.
O Christmas tree, O Christmas tree; Of all the trees most lovely.

O Christmas tree, O Christmas tree; Your leaves will teach me, also,
That hope and love and faithfulness; Are precious things I can possess.
O Christmas tree, O Christmas tree; Your leaves will teach me, also.

Edelweiss
Edelweiss, Edelweiss,
Ev'ry morning you greet me.

Small and white, clean and bright,
You look happy to meet me.

Blossom of snow, may you bloom and grow,
Bloom and grow forever.

Edelweiss, Edelweiss,
Bless my homeland forever.

Our Besharung was in the evening before Christmas day. It was the Christ child that brought the gifts. We only knew about St. Nicholas, nothing about Santa Claus.

On Christmas day we went to Christmas Mass, which always seemed well attended. In the very early 1920s we walked to church, the same church we go to now. It was only about a 40 minute walk.

Christmas was to all be home together with a fire burning in the wood heater, celebrating the anniversary of the Christ child and later having a nice roast dinner with mashed potatoes, gravy, salad, and canned beans from the cellar. All day Christmas we could look over and enjoy our presents, which were mostly things to wear and maybe a stocking with nuts, a few candies, and a few oranges.

Then, on the evening of Christmas day we would again light the candles, turn the kerosene lamp down, and sing Christmas songs for as long as the candles would safely burn. We would celebrate this old tradition every evening through New Year's Day. When the tree got too crisp to be safe, Mother and helpers would dismantle the tree and put the ornaments away.

After Dad got the Dodge touring car, we used to go over to North Aberdeen and visit Mr. Dinse, a German friend of the folks. He was an ex-sea captain and lived next to the North Aberdeen bridge. It was his job to open the bridge when a tugboat needed to get through.

The bridge was built of Douglas fir timbers. It must have been about 180 feet long and was balanced over the center turning mechanism. A tug would blow its horn for the bridge and Mr. Dinse would run out to the bridge. He would turn the horn on, close the gates on both ends, and then, with a long sweeper pole, turn the capstan back a little so he could unlock the lock dogs. Then he would walk around and around pushing the capstan lever, which activated the gears and started the bridge to open. When the bridge was at a right angle to the channel, it came to rest over the bridge gridiron, and the boat could get through with the overhead obstruction cleared out.

There were two mills remaining on the Wishkah, including a shingle mill on the east side of the river about one-and-a-half miles up river from the mouth. It belonged to Bill Rosenkrantz. Also, there was a spruce mill a half mile upriver from the mouth, run by Fred Hulbert, Sr., and owned by Sudden Christensen of San Francisco who must have gotten into the mill business after the San Francisco earthquake. The spruce mill used street car tracks for shipping to railhead any lumber they had to go out by rail.

Little tiny steam schooners with seven to eight feet of draft came up the river. The schooners were called "coasters," named so because they were in the coastal trade between Grays Harbor and San Francisco. They took the place of the various types of windjammers that had plied the water carrying Grays Harbor lumber since the 1890s.

American Mill No. 1, the spruce mill, seemed a big mill when we were children, but it must not have been. I know the mill and mill yard were small because Market Street was in the exact same location then as now and the land is only about 150 feet wide between Market and the river. The mill dock would berth two of the little coasters and could not have been more than 50 feet wide because the Wishkah there is not over 275 feet wide.

It was a steam mill. I don't know if it had line shafts or a turbine and generator which would have allowed the mill to operate with electric motors. All in all I don't believe there were more than four or five mills in the 1920s that operated with electric motors. Little did we know someday Marzell and I would build an all-electric modern mill.

There was no public electric power available for operating mills. The producer of power was a private company called Grays Harbor Railway & Light Co. Besides furnishing power for lights at their electric park power plant, they operated street car service to Hoquiam, Aberdeen and Cosmopolis. We used to ride the street car from North B Street to Heron and G for a five cent fare or catch the street car at Broadway and 4th and transfer at Heron and Broadway and on the same nickel ride to the end of B Street.

Grays Harbor Railway and Light headquarters were in the Electric Building; now the Failor Building. If you look closely at the building from Heron Street or Broadway, you can see hundreds of light bulb sockets into which light bulbs were screwed to light up the building. The architect that designed that building and whoever had the idea of the light bulbs probably got canned. How would you like to keep changing those bulbs?

As we got a little older our horizons expanded. Marzell and I read a lot. There were not many radios in those days, but our neighbor, Walter Haney, had a radio about 1927. We were allowed to go over on Sunday and listen to the radio music.

Mr. Haney bought a new 1928 Model A Ford coupe and he did not feel secure driving it so he asked Marzell to drive it and on days Mr. Haney needed to go to town, including to Sunday Mass, Marzell drove him.

Walter Haney was a real pioneer. He came to the Harbor in 1892 from New Brunswick, Canada. He was a logger all his life and worked

Walter Haney (on the right) at his shack in camp in 1902. He would not live with the other loggers.

on the Wishkah boom for Sam Ball. He usually worked at the sorting gap pushing logs at the first down-river log separation gap. He would let fir and hemlock go straight through and pole off cedar and spruce into a side race for later sorting.

From the porch of our home we could look out over the river and when the boom crew was rafting logs on the ebb tide we could watch them work. Our home being on a slight rise increased the view. We knew more than the boom crew would in their procedure of rafting logs. It especially intrigued me and in later years when we were logging into the river, I already knew the entire procedure.

During the peak years of the Donovan Corkery Logging Co., the boom crew rafted from three to four rafts every day, usually three rafts of fir and one raft of hemlock, cedar, or spruce. The boom crew foreman was Sam Ball, one of three brothers that came to the Harbor from eastern Canada. All of them worked on the boom. Most of the rest of the crew were from either Nova Scotia, New Brunswick, or Maine and some of them had worked on white water drives back east.

A pike-pole, the boomman's tool, has a jab point used for pushing and a 90-degree hook point for pulling. The pole is about 16 feet long. The boom crew had a double-ender gasoline powered launch with a little open-ended cabin on it. When coming down the river with all 12 boommen aboard, each standing holding his long wood-handled pike pole, all dressed in their own attire, you could tell they were geared to go to work.

Watching all that logging activity, it's no wonder we became loggers. I was always a tiny guy, and being a small person I could quickly jump over loose logs. Most loggers, especially the Scandinavians, were big men and that's probably the reason we went on our own. Nobody would hire a small man for cutting crew work. It turned out for the best. We had a good long-shot at running our own logging outfit. None of the big men I hired and worked with in later years ever said a word about such a small guy being a logger.

We had a little dock made of logs and planked over to which a boat could be moored. Also, it was in a little cove, safe from river traffic. We had our old red tub moored there and the log scalers had a real nice red boat they moored at our dock. We collected one dollar per month from the Scaling Bureau for that right.

We were about 12 years of age then. In those days there was no age limit on either end of life span. At 14, if you could do the job, you got hired.

As we got a little older, we ventured further and one of the most fascinating places was the Donovan Corkery Logging Co. camp and logging operations located at the Wishkah head works above the Aberdeen City water dam. Marzell and I went up there two different times. Once, when we were camping at the old Malinoski dam, we hiked about three miles up the road to where the loggers were building railroad grade right down parallel with the water line. This grade later became the road to the head works.

Road building was being done with a steam donkey pulling a slip scraper, dragging dirt out of the cut and making a fill. It was really something. In those days, 1927, there were no bulldozers. Many road building projects still used horses and Fresno scrapers. On a little one-man scraper you tripped with the handles. If the cut or fill being made was longer, a belly dump wagon was used and a steam shovel was used to load the wagon. It had to be dry weather in order to build road with that kind of equipment.

The following year, we walked a long way up the hill, beyond the city water dam and the camp, to where the crew was falling and bucking those great, big, old-growth Douglas fir. Seeing a steam yarder logging operation is something to remember, though we came years too late to have a logging show like we saw at the Wishkah headworks. The logs were laying two and three deep. The timber stood at 100,000 board feet per acre for a section and a half. No wonder we headed for the timber as soon as we were old enough.

After that timber was cut and logged out it was the end of Donovan Corkery. All the logging machinery was hauled out and sold, or, more likely, scrapped out and shipped to Japan along with the railroad rolling stock and rails. Donovan No. 1 mill closed, Donovan No. 2 mill ran through World War II, and then was also scrapped out.

Living on the Wishkah upriver from us at the Aberdeen Gardens were Mr. Adolph Bold and Mr. Willumsen. They both had double-ender gillnet boat painted green and every fall in the 1920s we would see them heading for the lower harbor to get in on the fall and early winter silver salmon run. Mr. Bold sometimes dropped off a couple of nice salmon at our place and mother would clean and can them.

Fishing was a major activity on Grays Harbor before the logging and saw milling. During the 1920s, the fish runs were still very substantial, and there were still several fish canneries in Aberdeen and Hoquiam, including Halferty Cannery at the foot of F Street. Every river, and even small side creeks like Stewart Creek, Fry Creek and Charlie Creek, supported good salmon runs. Apparently it was not so much logging the land off that caused depletion of the fabulous fish runs, as it was overfishing for so many years.

Marzell and I ran a small trap line in the late 1920s. One was on Stewart Creek. During December we would see lots of salmon in the creek already over a mile above tidewater. They were ready to spawn. Now days, I doubt if you could even find one salmon in Stewart Creek at spawning time.

We also had a trap line on the main Wishkah River. In the months of November, December and January, we would go out after school and check our trap line at low tide. The most common catch was muskrat. We caught lots of them out of their burrows along the river. We caught one or two mink, a raccoon, and lots of civet cats which stunk like hell and were hard to skin. We never did catch an otter or other exotic fur bearing animal like a fisher or a martin. Beaver trapping had been outlawed many years before.

Before 1929 we got up to $2.00 for a cured muskrat fur and I think the mink we caught sold for about $24.00. Later in 1930 and '31 fur prices dropped way down, muskrats down to 75 cents per hide.

On the river, we either checked traps upriver or down. We always had to buck the river current one way. We used to walk the slippery boom sticks with rubber hip boots. Had we ever fallen in, we would have had a time getting out, but it was better doing what we did then than all the trouble there is available for young children to get into today.

At the end of the boom or bust era, Aberdeen's "skyscrapers" were built; the old Daily World Building, the Becker Building, S. K. Kress Building, Sears-Roebuck Building, and the J. C. Penney Building. Most of them now hardly justify their existence financially.

In 1925, the Electric Light Co. brought power to the people living on the Wishkah Road. This was a great event. It meant no more kerosene or gasoline lamps to clean and maintain, though we still used the gasoline lantern for outside at night, to go to the barn or chicken coop or for walking up to the neighbors.

By 1929, the power line went clear above the Wishkah High School and up the East Hoquiam. It was a good thing power was brought in when it was, because after the crash of 1929 the electric company would have had a difficult time justifying such a project. Power would not have come in until cheap power became available from Bonneville, which was in the early 1940s.

About the same time, Mike Lentz headed up a group that set up a farmers' line telephone company with sufficient circuits that a much higher percentage of people could have telephone service.

It was a co-op venture between the telephone company and Wishkah residents. Mike and his brother-in-law, Mr. Paully, cut poles wherever they could find them along the road and pulled them out with a team of horses. After the poles were set up, telephone company people strung and mounted the galvanized telephone wires.

Telephones were good for the bootleg business. Booze distillers could quickly make handy drop-off arrangements. As youngsters we used to see a car stop and someone run into the woods lugging a sack, coming out in a few minutes. In a little while another car would stop, run in the woods, and come out with the sack of "goodies," put them in the car and take off. This was called a drop-off and pickup system, and worked so long as nobody got in between and high-jacked the goods.

Year by year there continued to be improvements. Roads were getting better and cars were getting fancier. The Wishkah Road was gravel all during those years. Sometimes in the summer the county would put some oil on the gravel to hold the dust down.

In 1927, Charles Lindberg flew all by himself from New York across the Atlantic to Paris nonstop in a modern single-wing plane built of aluminum, canvas, and Sitka spruce. The Sitka spruce came from Posey

Mfg. Co on Grays Harbor.

Later, Colonel Lindberg made a long flight with his plane, the "Spirit of St. Louis;" a barn-storming mission clear across the United States. He flew right over our house and over Hoquiam and Posey Mfg. Co., the place that manufactured the Sitka spruce lumber which went into the wings and framework of his plane.

Those were the fledgling beginnings of serious exploration of ways to make the skies a safe and speedy way for travel. About that same year a large dirigible named the Shenendoah flew clear across the United States and over our house on the same route as the Spirit of St. Louis had taken.

Grays Harbor was a famous lumber trading center during the 1920s. Aberdeen boasted the Grand Theatre to which any stage play, musical or opera that played in San Francisco usually was invited. While I was a first- or second-grader going to St. Rose Academy, the school had a play which was performed at the Grand, and I had a part in the play.

Working at Huffman's garden and other odd jobs I saved up a few dollars, maybe 25 or 30. Dad was banking in Hayes & Hayes Bank and our money was put in the same bank. The bank was under control of one Billy Patterson, the man instrumental in building the Grays Harbor Country Club. He lived in a mansion on the block where McDermot School now stands.

This story of Hayes & Hayes Bank was told to me by Dan McGillicuddy.

Billy Patterson called the shots, and he loaned money to a group that was in the logging business at Independence (near Centralia) to build a railroad into their timber holdings up Independence Creek. Dan was hired by some of the directors of the bank to go in and take a look at progress being made. They had already loaned a million dollars and it was not supposed to cost that much. 1927 dollars bought much more.

Dan told me, "When I got up there I was shocked. They had not finished even any part of the railroad."

When that news hit the directors, they tried desperately to shore up the bank's reserves and speed up collections. According to Pinky O'Day, they sent a runner out early before the bank opened in the morning so they could cover the day's cash needs. One day the runner didn't make it before the bank opened, and the State examiners were already there. The bank was declared insolvent by the examiners and closed.

Billy Patterson, until this fiasco, had lucked out on his seat-of-the-pants philosophy, but the Independence logging company guys took him and the bank for a ride. Instead of tending to business, they went on a safari to Africa and a whole bunch of other non-business activity, in reality robbing the bank. The bank was liquidated and paid out around $.65 on the dollar, which was a lot better than payout of banks after the 1929 stock market crash. Mr. Patterson was charged in court with malfeasance. He lost his home and ended as a broken man after many years as the lead banker of Grays Harbor and a community leader often in front of community projects.

President Harding died in office and was succeeded by Vice President Calvin Coolidge, and on August 2, 1927, Coolidge told the Republican convention, "I will not be a candidate for the President of the United States."

The Democrats picked Al Smith, one-time Governor of New York, as candidate. The Republicans picked Herbert Hoover, who had been appointed to do relief work in Europe after World War I. Hoover won the election and was sworn in as 31st President of the United States. The world was at peace, and the economy was booming along. The stock market was making the small ordinary investor rich. It seemed like everybody could get rich. In October, 1929 things changed. The stock market collapsed and dropped more than 50 percent of its pre-crash value.

Shortly after the stock market crash, another bank failed in Hoquiam, and the savings and loan could not pay out cash. They issued what was called savings and loan stock, worth in cash or trade about 50 cents on the dollar. Any deposit a person had made was redeemable in savings and loan stock.

This created a medium of exchange which was used for about five years. At that level the depositors in savings and loans were able to get some equity. We had a few dollars in the savings and loans, and we cashed some of our savings and loan stock. Thus began the Depression.

Chapter 3: The Depression

Two Men and a Horse

I GRADUATED FROM WEATHERWAX HIGH in June, 1932, with a class of 300. My grades were barely passing. Of mathematics and grammar I knew nothing. My mind was not on school work.

The summer of 1932 was very wet. It rained every day. Dad went broke in the shoe repair business. People had no money to pay for fixing their shoes, and Dad finally ran out of leather and had no more cash to buy another hide of sole leather.

Dad was renting space from Count Zelasko in the building on East Wishkah Street where Grays Harbor Transit bus station is today. Dad was not able to pay the rent so he made a deal with Count and traded the shoe repair machines for the rent due. Count took the machinery, and Dad brought home the foot treadle machine used for sewing uppers. I don't know what Count Zelasko did with the machines. They were about 1902 vintage because the original shop Dad bought was originally installed in 1904 in the Post Office block, at 105 South G Street, by Jeff Garman.

Fortunately, Dad had the home place paid for. When times were good in the '20s, he purchased some City of Munich, Germany, street improvement bonds, "non callable," which matured after he lost the shop. Cash from these bonds in hand, he asked Catherine Horning and her brother Lewellyn Graham, "How about giving me clear title for $1,200, hard cash?"

They snapped it right up and Dad got clear title on the home place two and a half miles north of Aberdeen, where Marzell and his son Thomas now live, 55 years later.

Dad had $85 left which he gave to Marzell and me. We traded

our little red pony named Bessy and the $85 for a draft horse named Maude.

Unless you lived then, you have no idea how hard it was to find any kind of work. We were young and eager to work but with no particular skills, and this was before the Civilian Conservation Corps (CCC). We used Maude to get out stove wood.

Fall came and Marzell was in 11th grade at Weatherwax High. I enrolled in Grays Harbor Junior College, in the old Franklin Grade School building between Market and First where the baseball field is now. The college had a wood-fired heating system and accepted wood for tuition. Marzell helped me get out wood in the evenings. We sold some to pay to have it hauled from across the road from where we lived to the school.

Most of the students at the college, including Jim Quigg, Bob Harriage, Merle Schmid and Lawrence Warwick, were very intelligent. They could play bridge during study period and still get their work done. I was lost.

I was just wasting my time, so in January, 1933, I gave up the college, though I had another quarter of tuition paid. Dean Tidball, who was running the school said, "You will be sorry later that you did not finish and stay with getting a college degree." He was a fine man, and I finally was sorry but it was fifty years before I began to feel that way.

Marzell and Hedwig continued school. I worked with a logger named Martin Foshaug getting out wood further up the road from the home place. We usually cut remaining old-growth fir, snags and windfalls. Most of the land was reverting to the county for back taxes. Anyone that could squeeze a hard cash dollar off it was not bothered.

We had an old Vaughn gasoline-powered drag saw for sawing the logs into 16- and 24-inch lengths. We sold the wood for $3.00 a four- by four- by eight-foot ricked-up cord, delivered. We advertised in the Aberdeen Daily World as "bone dry forest wood." Some of it required a blow torch to get it to burn, but people were so nice.

It was the worst of times. There were strong, young, able-bodied people on bread lines. Homeless, hungry people marched on our national Capital. This situation began under President Herbert Hoover and had already gone on much too long. On March 4, 1933, a new President, a Democrat named Franklin D. Roosevelt, was sworn in. About the same time, two young men took office who would become the best Senators that ever represented our state in Congress; Warren Magnuson and Henry Jackson.

Politics were affecting countries world-wide. Hindenburg, President of Germany, died. Adolph Hitler jumped in and got control of the destiny of that country. Mussolini took control of Italy from their king, Emanuel. The Emperor of Japan let Tojo take control of Japan. Stalin, most infamous of all despots, cut and shot his way into power in Russia. Great Britain and France were status quo, and here in the United States people went about their daily lives paying little heed to the rumblings in Europe.

Summer came and we helped at home with the berry patches and the garden vegetables. Medication derived form the bark of cascara was used by the medical profession as a physic and laxative and we peeled it in June and July and used the horse wherever we could to get the bark out. We dried it on the barn roof, and on wires and then cracked it, stacked it and stored it in burlap sacks stacked in the wood shed.

In August, 1933, we got our first logging contract from Mr. J. R. Thomas, who was operating a pulp wood show just south of Cosmopolis. Our job was to build skid trails and load the wood on a 12-foot-long sled which the horse Maude pulled down to the loading area. George McKay, using his brother Art's Ford truck, hauled it to Grays Harbor Pulp & Paper Co. in Hoquiam. We completed skidding out all the wood at the Cosmopolis show, and Mr. Thomas had us move to another little pulp show he had at the foot of Evans Street in Aberdeen.

Working with horses was very difficult during the rainy season. Our clay soils become a sea of mud which, when the rain stops, becomes sticky gumbo. We had to wash the horse's fetlocks every night or the wet mud would cake under belly and fetlock and hair would come out. We had to be careful not to hurt Maude.

In December, 1933, there came a southwestern rain storm the likes of which had not been seen around the Harbor for some time. It rained over 31 inches. All of South Aberdeen was flooded.

We could no longer get wood out for Mr. Thomas. The only way I could get to our horse, stabled in a barn at the foot of Evans Street, was on foot through Cosmopolis on Curtis and West Boulevards, then up past the riding academy, through an overland trail and to the foot of Evans Street, which was five feet under water. I led the horse through, over the trail and then rode her through Cosmopolis, South Aberdeen, across the West Bridge, across the old Heron Street Bridge and home up the Wishkah. We never went back to that job.

In late December, we got a job with Earl Karshner, a pulp wood contractor working on the Martin Wood place about three-fourths of a mile by boat down the Wishkah River from home on the Bear Gulch side. He had fired the Linns, who were trying to do the job with two mules.

Mules do not work well in soft mud. Their feet are too small.

The week between Christmas and New Year we made $32 hard cash. We went to work in the row boat, almost always bucking the tide coming or going. The only expense was one sack of rolled oats at $1.05 for Maude and a bale of hay a week at $1.35.

The Wood's timber tract was a fairly level plateau on top of a hill with steep slopes running down to the county road. Mr. Karshner had a Mr. Weaver build pole chutes which carried the wood from three strategic points down to a truck loading area. It was a well laid out pulp wood show and the weather turned nice. We were able to work steady and pay our bills.

Hedwig finished Weatherwax in 1934 and Marzell and I were able to help her enroll in Central Normal School at Ellensburg in the fall of 1934.

We finished the M. Wood logging show for Mr. Karshner and for part of the winter we hauled out wood for Charlie Cyr, another pulp contractor. In the meantime, Mr. Karshner bought a fine stand of mixed old- and second-growth hemlock from B. Horning & L. Graham, 40 acres of which was located just across the road from our home place, for about $1,200.00. We started that job in the dead of winter and fought it through until dryer weather came. We were able to use an elaborate skid road system for the wood sled.

Roosevelt, our new president, shook up a lethargic Congress and things started to happen. In 1936, Fort Peck Dam dammed the Missouri river and is still an effective flood control dam in Eastern Montana. Also built were Grand Coulee across the Columbia above Wenatchee, and the Bonneville Dam on the Columbia just above tidewater.

The three C's were set up and young men 18 and over were able to get a job doing honest work and a chance to feel good about themselves. Pay was $1.00 per day and food, clothes and medical.

That summer of '35 we increased our horse power by the purchase of Jerry, a bald-faced bay with three white stockings. We worked him with Maude. He always let Maude start the load. Then, he would prance and pull like hell. That fall we bought a beautiful dapple grey 3-year old gelding from Mr. Crass on the Wynooche, and named him "Bob."

Mr. Joe Silva helped us by then. He was a good man with horses. He grew up in Talent, Oregon. Jake Mayer, an old time German friend of Dad's, worked part time for us building skid trails. Mr. Karshner had two trucks which we had to keep in wood. We also took care of the pulpwood cutters and all the layout for him.

The winter of 1935-36 came. After New Year, we finished all the

Marzell Mayr and Jacob Mayer with "Fanny" and "Bob." Fanny "cost $200 and we had to bury her when she died."

area that could be worked with horses. The new area was too steep for horses. We moved into eight-foot alder logs for Heine Anderson's alder mill at Port Dock in Aberdeen. A man named Davidson was in charge.

Heine Anderson was good to work for and a tough fighter. More than once he lost valuable alder lumber because it was green and in shipping it to Los Angeles it was in the hold too long and got red stain. Alder was sawed into one- to four-inch planks and shipped rough and green by boat to Los Angeles where it was used in furniture manufacturing.

We said "good-bye" to the back-breaking pulp wood forever and in a few years the Grays Harbor Pulp & Paper installed a drum barker and 120-inch chipper. Schafer Bros. and Polson's Eureka mill kept the mill in chips, and pulp wood logging became history on the Harbor.

In January, we obtained a small patch of alder from A. J. Stewart. His mother, Jean Stewart, came from Aberdeen, Scotland with her husband around 1880. They came by boat to Samuel Benn's place on the Wishkah River. When it came time for the town to be named, it was Mrs. Stewart who was asked for a name, and she named it after her home in Scotland.

The alder patch was located across the Wishkah River from home. We paid Mr. Stewart $20 hard cash. We rolled the eight-foot alder logs to the river bank by hand, brailed them together into a little boom, and floated them down to the old school house where there was a little loading platform Si and Hiram Hulet had built a few months before when they floated logs down the river from clear above tidewater. We hired someone to haul the logs to the alder mill. From there on we expanded rapidly.

We obtained a four-year-old iron-grey mare named Fanny. She was a powerful horse, part Clydesdale, but she got sick that February while we were getting out alder on the Larkin place, now the Neil Tikka place. That beautiful horse cost $200 and we had to bury her when she died.

Finally spring came and dryer weather. Together with that and a new Federal program called NRA (National Recovery Act), things went well during the summer of 1936. We logged alder wherever we could find a patch that could be worked with horses. In July we were on Elkhorn Creek near Raymond before moving up to Satsop near Matlock and on to the east fork of Wishkah.

We got some nice alder on the Baleville Road, near Raymond owned by a Mr. Davies. By then we had a one-and-one-half-ton 1935 Ford flatbed we could haul 1,100 to 1,200 board feet on, Doyle scale.

Mr. Tully Stallard was office manager for Heine Anderson. In the late summer of 1936, we needed more horse power and Marzell told Mr. Stallard that we needed $350 to buy another team so we could rotate the horses. Mr. Stallard approached Heine, and Heine said "If they need a team, they need them. Tell them to go ahead, but I don't know where we will get the money."

We bought a team of bays and used them alternately with our other team, two weeks on and two weeks off. Anderson advanced us the money because he needed our logs.

That year the big forest fires on the Oregon coast were burning. The sky was grey from smoke and ash, but it did not begin to rain until December that year. Dad worked with us that winter until we finished all we could get with horses and we left Raymond and logged a few patches around home and fought the winter elements again.

In the spring of 1937 we obtained a two-axle-drive Ford, and the alder market went down from $14.00 to as low as $11.00. We even logged cottonwood in eight-foot lengths off land located on the South Shore of the Quinault River, about three miles above the lake, just above Norwood Guard Station. We needed to peel off the bark so the horses could pull the log in the sandy river bottom. The summer went fast but the whole economy went down again in 1937.

The "tough guys," Hitler, Mussolini and Tojo, were flexing their muscle. Hitler saw that the German people were not ready for a democratic government. When Hindenburg died, he jumped into the void.

Germany had a large population of Jewish people, well-educated and in many of the professions. Hitler stirred up hatred and less than four years later, his system lashed at the Jewish people with what was known as the Crystal Night Pogrom (November 9 - 10, 1938), just the beginning of his "final solution." Some of these people were able to get out and come to America before that time. Some came to Aberdeen and Mom and Dad visited them. It eased their pain and homesickness. We spoke fluent German.

In the winter of 1937-38 we made a decision to go "long logging." We sold the horses, and bought a tractor and the bare necessities, which included a truck, bunk and log trailer with accessories that cost $1,350.

We obtained our first tract of hemlock in the Newskah from Barney Stout and during the summer of 1937 laid a 1,500-foot-long plank road all by hand, back to the foot of the hill where the timber stood.

There was only one market for hemlock logs; Grays Harbor Pulp & Paper. I went to their office twice every month from May until November and asked Mr. Otis Hallin, log buyer, if he would buy some hemlock logs.

In November, 1938, he said he would buy our logs. This coincided well with the rainy season. We barely got started and again it was gumbo mud so deep and gummy we could not get the tractor off the plank road.

We loaded logs with a gin pole (called such because it leaned over like a drunk) on which the loading block hung over the center of the truck. Gene Carlson was our cat operator. Backing up the little BD Cletrac, Gene slewed the logs over the truck, and, with one of us on each end, Marzell and I pushed the log in place on the bunk.

It was December, 1938, and on some days we worked into the dark, signaling Gene with a flashlight. We dumped into the Chehalis River about where the Morrison Park is located in Aberdeen.

It was a railroad log dump converted to a truck dump owned by Vic Morrison. We sold our logs with Saginaw Timber Co., who had their office on the 5th floor of the Finch Building. Bill Morley was manager. We sold the logs for $9/M (thousand board feet) Scribner, less dumping, booming and rafting at 75 cents/M. Saginaw Timber Co. took a small commission.

God blessed us with the patience and courage to continue in the face of nearly insurmountable odds. We drew no salary. Helping on the home

Mayr Bros. No. 1 and No 2. These trucks were often loaded in the dark.

place paid for our food and room and Mother washed our clothes, and it was many years before we could repay our beloved parents for all they did for us.

In the meantime, Hedwig got her teaching certificate and her first teaching job at Naselle, Washington.

Our sister Margaret went directly to St. Joseph Hospital from the ninth grade. There she studied under fine doctors, including Dr. Kinne, Dr. Brachvogel, Dr. Graham and Dr. Goodnow. Margaret obtained a nursing certificate and continuously served the people of Aberdeen for over 20 years. Later she moved back to Columbus, Ohio, where she first came to live with Mother and the rest of us on arrival from Switzerland.

Margaret followed the nursing profession until her 70th birthday, mostly in children's hospitals, at Columbus, Ohio, and Little Rock, Arkansas, and after her retirement spent many more years as a volunteer in Little Rock counseling and mentally preparing children for surgery. Margaret lives today in Aberdeen Manor.

Marzell and I and our partnership were gaining the respect of the forest products and logging community. Malcom Stewart, another son of pioneer Jean B. Stewart, helped us make acquaintance with Mr. Arnold Brandis, trustee in the liquidation of Donavon Corkery Logging Co., who had been logging on the divide between the Wishkah and Wynooche Rivers clear up to the Aberdeen headworks. We wanted to buy the remaining timber on the holdings he was liquidating.

I went to meet Mr. Brandis in the Northwest Logging Co. office which was about where Swanson Hoquiam Market is now. I found him to be a fine gentleman, and we bought the timber.

The remaining patches of nice hemlock, a few old-growth fir and spruce were located from Perkin's Eddy upriver to the Donavon Corkery log dump. It was a thrifty operation. We again went to work in our trusty row boat.

Getting the tractor from a road down to the logging job was something else, but we did it. We went down the Donavon Corkery railroad grade and Gene Carlson drove the Cletrac across an old trestle 60 feet off the ground.

We put the logs in the water at several points of high ground along the river. We rolled them over the bluff at Perkin's Eddy and just above the Donavon Corkery log dump. Mr. R. J. Ultican, Sr. sent us a couple sets of boom sticks.

We rafted the logs with the help of Martin Foshaug. However, before the tug would pull out the raft we had to go at low tide and cut off any piling that restricted the channel for towing. That was a dirty job and took a few days.

We had a good relationship with Grays Harbor Pulp & Paper Co. Mr. Otis Hallin, their buyer, scaled the logs at the pulp mill. I had also scaled them before they were towed and Marzell and I thought I had made enough deductions, but I guess the logs got thinner on their way down to Hoquiam.

A boom crew consisted of six or more men who stowed logs in the boom sticks, and three or four men who worked above at the sorting gap. Larger logging operations could make two sorts simultaneously and the crew worked like hell, dancing across the logs like jack rabbits. They knew they had only six hours of ebb tide and better damn-well get all the logs in that day's run stowed and swifters across so the foreman could attach the raft number.

The number was phoned in to headquarters and became basis for sale of the logs. The log scaling was done by Grays Harbor Log Scaling and Grading Bureau. Formed in 1924, the Bureau gave the industry a good set of scaling and grading standards which have been in continuous use by the entire forest products industry on Grays and Willapa Harbors since.

The year was 1938. I had just turned 24. Marzell was 22. Wesley Foshaug, a friend from the Wishkah was 16 and worked once in a while. Gene was 29. Dad still worked with us once in a while. The labor unions were getting the entire logging, saw mill and plywood industry organized.

Grays Harbor area was graced with some unique union leaders, including Ted Dockter, Dick Law, Denee Dyer and Ted Cochenette. Some said they were communists. Of course we did not know what a communist looked like, but I thought they were different.

Since the very first trees were cut into logs and dumped into any of the main tidal streams of Grays and Willapa Harbors, booming and rafting grounds were designed so that logs floated out on the ebb tide. Tides are six hours out and six hours in. Boommen and rafters, with their own union, were able to negotiate a six-hour day because their work day was set by the tide.

We were a tiny fledgling outfit but by the grace of God, Buck Buchanen happened to come up the river by us on the Ultican tug "Hustler." At the very next meeting of the boommen's union, he reported an outfit six miles up the Wishkah River putting up log rafts. The result was "sign 'em up or picket the raft." Denee Dyer, business agent for the union, put the heat on us and Ed Hannick picketed the finished raft in a row boat.

We told them to go jump a stump and even tried to pick a fight with two of them on the river bank above the Pest House but they were well coached, knowing a fight would kill their mission to organize one more little outfit.

I went to work. The G. H. Pulp Co. said they would buy the logs if we got them to the mill, so I found Cap "Scovey" Mercich who had a little tug named "Tillicum," an Indian word meaning "sweetheart" or "loved one." She was a World War I sub tender, iron hulled, and had spent considerable of its time on Grays Harbor underwater. Scovey said he would tow the raft, picket or not.

I got worried and checked with one of the most brilliant attorneys the Harbor ever had, Mr. W. H. Abel of Montesano. With a deep drawl, he said, "Yes Werner, I can get you fellows an injunction. It will cost you $150.00 up front."

I was still wary and went to see my friend Mr. Dick Ultican, Sr., and laid the story out for him.

He said, "Werner, you are on Grays Harbor, the throne of organized labor. You have two ways you can go. One, fight with organized labor, and that's what you will be doing, or sign up with them and proceed with the logging. Better times are coming."

Marzell and I discussed this and made the decision to sign up. We paid dues on three people. It cost all the money we had in the bank, $15.00. The boommen let us use one of their men who worked in the woods part time until he got a better job. Later, we used George Debuque, a professional boomman living in Cosmopolis.

The crew in 1939. From left, Joe Zembal, Frank Hannick, Gene Carlson, Marzell, and Wesley Foshaug. I was running the camera.

In the early days we had no problems with the union except when there was a general strike. Then, we could not work either.

Organized labor did not understand how hard up the industry was on Grays and Willapa Harbors in those early days coming out of the Great Depression. The hard times on the Harbors did not end until years later, after World War II. They lasted longer on the Harbors because the sawmills cut out of timber supply at about the beginning of the Depression.

We got through the winter of 1938-'39 and bought a brand-new DDH Cletrac tractor from Elmer Schoen, of Pacific Hoist & Derrick in Seattle, for use in yarding logs. It cost $8,500.

Later, Mr. Schoen told me a story.

After the sale, he walked into our bank, National Bank of Commerce, and offered them the paper. Bob Crook, a bank officer, said, "Why did you sell them a tractor? They are having a hard enough time as it is meeting their obligations."

Mr. Schoen explained nothing to the bankers because he could see they were too stupid to see a potentially good, sound account with a great future.

Elmer Schoen reported to his father in Seattle, owner of the company that sold us the tractor, about the bank's unwillingness to buy the

Logging in the bed of Finley Creek, January 1940.

paper. He said, "Elmer, what do you think?"

Elmer said "I have been following their operations for two years and know their parents. They have good business sense and lots of work opportunity."

Mr. Schoen said, "That's good enough for me; we will carry the paper."

We were asked by Mr. Hallin of Grays Harbor Pulp & Paper if we would finish logging some felled logs on Finley Creek above Lake Quinault. We looked at the job. All together there must have been about a million board feet, mostly hemlock. We made arrangements to haul the logs to the Polson railroad reload just north of the North Shore Road near Quinault Lake for shipment to tidewater.

Gene Carlson ran the Cletrac; Marzell drove truck; Frank Hannick was hook tender and choker setter; and Wesley Foshaug and I felled the remaining timber. It was a good winter logging show except on the days when Finley Creek went on a rampage. Spring came in 1939, and we finished most of the timber we contracted for.

In 1939, we negotiated a timber purchase agreement with Mr. C. Davis Weyerhaeuser and E. Murnen of Weyerhaeuser Timber Co. on a tract located on the road to Raymond near Arctic in Section 31. It was supposed to be a thinning of 90-year-old hemlock. The stumpage rate for hemlock was $1.50/M, for spruce, fir and cedar, $3.00/M.

We built a plank road from Highway 101 to the foot of the hill. The weather was dry, and we worked there until the fall rains came. When we could no longer work the steep ground, we went back up to the North

Shore of Quinault and spent the winter of 1939-'40 finishing the remaining standing timber on Finley Creek. We went up the gravel bar with the truck and logged and loaded logs right in the streambed. Most of the time the creek ran underground.

I met George Kellogg, a fine gentleman and trustee in liquidation of E. K. Wood Lumber Co. He had a little office somewhere near where Major Line Cabinet is located in Hoquiam. The E. K. Wood mill was all removed by then, but Mr. Kellogg said they had considerable holdings on the north side of the Quinault River above Big Creek and that they would like to get it cleaned up.

He had made a timber sale to a logger named Lewellyn, a stumpage contract with set price for old-growth fir, No. 1, 2 and 3 peelers and a lower price for No. 2 and 3 saw logs. Mr. Kellogg said the logger felled all the big, clear Douglas fir but only logged out the first and second log and he wanted no more of that.

We looked it over carefully and, hell, they hadn't even taken out half of the peelers, and most of the fir was already bucked. With the down logs and the standing timber across Big Creek, there must have been somewhere between six and eight million board feet, and then, if a hemlock would not make a 40-foot, 18-inch top, we didn't even cut it. Also, we spring-boarded up sometimes two boards high.

We were successful in negotiating a contract with Mr. George Kellogg. No deposit, no bond, stumpage for hemlock; $1/M, spruce, fir and cedar; $3/M. We had a job for the winter of 1940-'41 and part of 1942.

While this was going on, we went back to the Weyerhaeuser section and built more plank road. All the truck logging in those days was on plank road. Hobi Brothers, with their North River Logging Co., used plank road exclusively for many years. Portable tie mills cut 4-inch by 12-inch by 10-foot planks and sold them to truck loggers at $11/M, lumber scale.

We purchased our first bulldozer, a DDH Cletrac, and that made road building much faster. We bought it from Elmer Schoen, our good friend at Pacific Hoist & Derrick. This time, when he went into our bank and offered them the paper, it didn't take five minutes. They snapped it right up like a fish taking a fly.

The logs from the Weyerhaeuser logging site were hauled to Blue Slough log dump and rafted along with those from other loggers and sold to the same mills they sold to. However, if we had 75,000 or 100,000 board feet of one particular sort, we could get them sorted and rafted separately for our specific customer. When winter came, we pulled out of Section 31 back to Quinault to start work on the E. K. Wood holdings.

The only logs we hauled to town were the Douglas fir peelers and they went direct to Old Aberdeen Plywood. Prices then were No. 1 peeler, $18/M; No. 2 peeler, $16/M; No 3 peeler, $12/M. No. 2 fir saw logs were $12/M; No. 3, $8/M. Hemlock was still $11. At Polson Railroad, transfer to rail car, freight to log dump, booming, and rafting totalled $3.75 per M, which Polson Logging Co. deducted from the invoice. We received a copy of the invoice on any raft we had logs in.

With two tractors and a small loading machine, we were able to get good production, up to twelve loads per day. New people joined our crew, including Joe and Jim Zembal and John Rycz. For these people it was the first job they ever had. Horace Alwood also came to work as truck driver.

On Valentine's Day, 1941, I met my future sweetheart, a beautiful blonde girl from Cosmopolis. She lived with her widowed mother. Jennie and I would meet Saturday evenings at the Dreamland, a public dance hall in Hoquiam where young people danced to such old timers as the Missouri Waltz and lots of Polka and Shoddish. There was no booze in those days, yet.

Jennie's mother was Brita Strindberg from the flat-land farming country of Ostersund, Sweden. Brita had 12 brothers and sisters. Her brother Gunnar and two sisters, Elsa and Jennie, left Sweden with Brita for America around 1910, coming all the way to Seattle. Elsa and Brita

Jennie Brolin Mayr. "On Valentine's Day, 1941, I met my future sweetheart, a beautiful blonde girl from Cosmopolis."

Above, Grandmother Emma Strindberg. Below, Jennie Strindberg, Jennie Brolin's aunt, died in Alaska.

worked as domestic household help in Seattle. Brita worked for people named Miller whose daughter married an East Indian prince and moved to India. Brita learned the English language from the Millers.

The Strindbergs were also musicians. Some of Jennie's cousins are musicians. Her grandmother was Emma Strindberg.

Jennie's father, Helmer, came from central Sweden, the flat farming country south of Strumsund. His father, Helmer Gustav Brolin, was a tailor. Grandfather Strindberg was also a tailor. Helmer and his brother Gustav came to

America from Sweden about the same period as the Strindberg family, just before World War I. They did not know each other until they came to Seattle.

Jennie and Gunnar Strindberg headed for Alaska. Jennie died as a young woman, contracting an infection after appendix surgery while living somewhere in Alaska.

Gunnar, who I met when we were young, was a tall, very handsome blond Swede. While in Alaska he met and married an Aleut girl. They had two children, both girls. Gunnar was living at Dutch Harbor, Alaska, in 1941 at the time the Japanese

Gunnar and Esther Strindberg, in Dutch Harbor. She recovered from tuberculosis. He was lost in Alaska.

bombed Dutch Harbor. The two girls and their mother were evacuated down to Seattle.

While in Seattle, their mother contracted tuberculosis and entered a sanitarium for treatment. We met her while she was in the sanitarium. She recovered and went to work in Seattle.

The two girls went to live with Aunt Elsa, who had married a timber faller, Carl Strom and lived at Monroe, Washington. We went up to visit them when we were first married. They raised seven children plus the two girls from Gunnar. Gunnar has not been heard from for many years. He followed fishing boats out of Alaska.

There were many Swedes and Norwegians on the Harbor during the years 1900-1941. Aberdeen and Hoquiam each had a Vasa Hall. During their lodge meetings they used the language of their newly adopted country, but when visiting with each other they spoke their native tongue.

A steam yarder pulling trees to the salt chuck, Wishkah River, in 1917.

On Grays Harbor, the Swedes went to work in the logging camps or the new mills. It was a warm, hospitable place for Scandinavians to live and work. They were well educated, had a sense of humor and were built powerfully enough for the heavy work as timber cutters, or in the mills dragging the heavy green fir planks and timbers off the green chain. Copenhagen chewing tobacco became an indispensable item for loggers and mill people. The very name denotes Scandinavia.

The Scandinavians worked for outfits like Carlson Logging Co., Carlson & Callow and Wynooche Timber Co. Also there were "Dirty Old Saginaw Logging Co.," Clemons Logging Co., Schafer Bros. Logging Co. and C.C.L. and T. Logging Co., and, of course, the greatest of all, Polson Logging Co., called "Papa Polson" in the trade. All had good Swede foremen.

There was fire in the woods with the wood-burning steam yarders puffing and yanking the big yellow fir logs to the skid road and the yarder pulling sometimes a mile on a skid road to the salt chuck.

Brita and Helmer Brolin, Jennie's parents.

Alice Brolin Walker, 1918 - 1941.

Some of the immigrants acquired a "stern picker," a fishing boat used in the rivers that set and hauled in its nets over the back end. That was hard work. You were either pulling the boat backward or the net forward.

It was a two man job but when fish weren't running good it only paid for one man, maybe only getting three or four fish per drift. Fish came in schools and when a real "run" was on they could fill the little boat to the gunwales. That may explain how Edmond Ness, a Norwegian logger, lost his life drift net fishing.

Brita Strindberg met Helmer Brolin in Seattle. They fell in love, were married, and moved to Aberdeen where Jennie and sister Alice were born. Helmer was a sawmill worker on the Harbor, working at Grays Harbor Commercial Mill in Cosmopolis and, after they closed down, worked at Anderson Middleton and later at Old Northwest Mill in Hoquiam. At that position he was mortally wounded in the stomach when a saw came apart and hit him. In those days, around 1935, there were no antibiotics to fight infection so gangrene set in and he died at 42 years of age.

Jennie's older sister, Alice, was married to Angus Walker. She was 22 years old when Jennie and I met and a very beautiful girl. I never did get to meet her. On a Friday after coming in from camp, I went to see Jennie. No one was home so I went across the street to her aunt, Mrs. Ohlinder. She told me something happened to Alice, who lived at Astoria, Oregon.

Next day was Saturday so I immediately set out for Astoria and found both Jennie and her mother and found out Alice had passed away from a self-inflicted gunshot wound. It was a very sad situation. After the funeral both Jennie and her mother rode back to Cosmopolis with me.

I loved Jennie and asked her to marry me. She finally did say yes and we were married in St. Mary's Church June 28, 1941, in a beautiful ceremony. We were both simple people with humble beginnings and that is the best. We set up our home in a little apartment upstairs in her mother's home.

The situation in Europe was heating up. Hitler was on the march, east toward Russia and west into France. Our President urged Congress to get moving on upgrading our military preparedness.

We moved back to Weyerhaeuser Section 31 and built a dirt road over top of the hill toward the headwaters of Charlie Creek. It was too steep for a plank road. We found some fine high-quality cedar near the west line.

Even though property lines were carefully surveyed and marked, Mr. Neil Cooney, owner of Section 30 to the north, found we cut one lousy hemlock tree on his side of the line. He sent us a bill for $1.50. He had around three or four parts of sections that had never been logged and many years later he sold them to St. Regis Paper Co.

The State had Section 36 adjacent to Section 31 on which stood a fine stand of old-growth Douglas fir. We applied for a timber sale with the state land office. Later that year the timber came up for open bid on the Grays Harbor County Courthouse steps. I went to the sale and bought our first State timber sale, a lump sum sale. It was high-quality timber, around four to five feet on the butt, with five-log trees that yielded two, sometimes three peeler logs.

We were logging up to 600,000 board feet per month with a crew of up to eight men. Logging was from stump to dump, including falling, marking and bucking, and all cutting was done with axe, springboard and cross cut saw.

We had good weather that year and were able to work into the fall. We finished Weyerhaeuser and the State timber sale, and moved our outfit back up to the north shore of Lake Quinault. We were getting quite a bit of medium-sized old-growth spruce which also went to the Polson Logging Co. reload. It was sorted out at their rafting grounds for their Eureka Mill, which stood far out on piling about where ITT Rayonier pulp wood storage is, east of their wood room.

One day, Arnold Polson called and said, "Our mill burned to the ground so we don't know what we can do with your and our spruce

Jennie and Werner Mayr, June 28, 1941.

Just married: Jennie and I depart for our honeymoon, June 28, 1941.

logs." Some weeks later, they acquired the closed Northwest Mill just under and below the Eighth Street Bridge in Hoquiam. There, they cut millions of feet of spruce from the Quinault country.

We produced Sitka spruce and hemlock logs and lesser volumes of Douglas fir throughout the summer and into the fall of 1941.

Most of our crew was in their early twenties and not married. We rented the Kestner homestead house, a large ranch-type home on the North Shore across Canoe Creek from the county road. The logging crew stayed there through two winters.

It was the only place in that country that had electricity at that time. It had its own little hydro plant which was only used in the night time. It had a unique system used to turn the water off or on, similar to the old steam donkey jerk wire. A set of two braided wires went a distance of 800 feet to the foot of the hill to the hydro plant water valve and around a pulley about 12 inches in diameter at the house. By pulling the lower wire, the water valve opened and allowed water to go through the turbine and spin the electric generator. Pulling the upper wire shut off the water and stopped the generator.

In August of 1941, the U. S. Forest Service offered a timber sale on the South Shore of the Quinault Valley above Lake Quinault about one mile below Fletcher Creek. Forest Service personnel at that period were H. J. Andrews, Regional Forester; Carl Neal, Forest Supervisor at Olympia and Joe Fulton, District Ranger at Quinault.

The sale was one of the first made in the Quinault Ranger District.

Volume was three million board feet of Sitka spruce, 500,000 board feet of Douglas fir and two million board feet of hemlock. The sale was by sealed bid to be opened at the Supervisor's office in Olympia. We bid $8.00/M for spruce, $8.50/M for Doug fir and $1.50/M for the hemlock. As high bidder, we were awarded the sale. We outbid Esses Logging Co. by 50 cents on the fir. The entire sale was "selective cut" logging. These logs also went to the Polson Logging Co. reload.

By then we had a FD Cletrac and Carco arch for cat logging. We built the necessary roads as quickly as possible. Dan Boone of Hoquiam did the road ballasting, digging right off the Quinault River bar.

Jennie and I rented a little house on the south side of lake and lived there and went home on weekends. Built as a sort of summer place, when the rainy season came, it was too miserable, so we moved back to our apartment in Cosmopolis. Most of the time I stayed at the batch camp, and went home on weekends.

At that period we had about twelve people in our outfit and three trucks. Louie Raich drove the E.Q. Mack 1940 model, Horace drove a new 1941 Ford Fabco, Harry Fry drove the third truck.

On the global front, nations were beginning to take positions. A person got an ominous feeling a terrible world-wide war was not too far off. So far, our President and Congress had helped Great Britain with some old, beat-up ships useful in shipping, some old armed vessels, and we were also providing supplies.

On Sunday, December 7, 1941, Jennie heard the news in the kitchen of our Cosmopolis apartment. As our President spoke of the terrible act the Imperial forces of Japan had loosed on our facilities at Pearl Harbor, Hawaii, Congress responded. The Japanese and the United States were at war.

Chapter 4: The War and After

From Horses to Tractors, And On To High-Lead

THE UNITED STATES WAS NOT PREPARED for a world-wide battlefield, but Congress and the President acted quickly. Seven of our young men either volunteered or were called up. Horace Alwood signed up in the Air Corps and went to the South Pacific. Horace was Canadian by birth but received U. S. citizenship at volunteering for service. Joe Zembal was in the African Theater and Italy. Wesley Foshaug was in the Pacific Theater doing his part in the Army. John Rycz served the Navy in the North Pacific. Julian Zembal was in the European Theater. I was never called up.

The government set up selective service boards, war-time ration boards, man-power boards. Each board had priorities. Selective service boards set up deferments, usually six months at a time. The system tied an employee to the company he was working for and provided a stable work force. I took care of this paperwork.

During the war years, running a logging and trucking operation presented problems never encountered before by us. Food stuffs like sugar, meats, eggs and other staples were rationed. No new automobiles were manufactured. Everything was for the war effort. Nonessential activity pretty much came to a stop until the winter of 1945.

Grays Harbor spruce lumber of airplane quality was shipped to Kansas and used in the manufacture of training planes for young pilots. There was an acute shortage of aluminum and all of the metal available was needed for bombers and fighter planes for the United

A load of 8,500 board feet of Sitka spruce on its way to war production.

States and our allies.

Our small company was a major producer of Sitka spruce logs which were manufactured into lumber at the Northwest Mill in Hoquiam. Beside the spruce, we were producing old-growth Douglas fir logs in long lengths which were cut into boat and barge lumber at the Grays Harbor Lumber Co., owned and operated by Nels Blagen.

We needed a tractor to log spruce off the Queets River drainage, but all new equipment was commandeered by the military. In 1942, I went to the Boeing Airplane Co. Seattle office, explained our need, and we were able to get an A-1 priority certificate to obtain the necessary equipment. They issued the certificate then and there, and I went to the equipment dealer, who immediately released a tractor to our spruce operation.

Loggers work hard and consume a large amount of calories, and from 1941 through 1945, we ran a logging camp and cookhouse. With food ration stamps, I was always able to obtain the necessary supplies for the cookhouse.

Logging truck tires were also rationed. Because we used the shortest haul route possible from the logging area to the Polson Logging Co. railroad reload, we were able to obtain sufficient ration stamps to keep our

A load of Spruce on Marzell's bridge across the Queets.

trucks going on a regular schedule. We were also able to obtain the necessary diesel fuel and oils and grease for our tractor.

Log prices were frozen all through the war, but for every four hours we could log over 40 hours per week, we received another 50 cents per M. Mills could add on, and collect from the buyer by presentation of the proper documents, additional costs from producing special orders which the buyer would ask for in the purchase agreement.

Production board freezes on log prices backed into stumpage. It was illegal to pay higher than ceiling price, so if all timber bids were at ceiling, the first option of the bidders was to cut and deal and winner took the timber contract. The winner might share logging or trucking with others, and agree to sell the logs to the various mills who had an interest and need for logs off the sale.

People were forced to toe a narrow line in keeping costs in line and lumber and plywood were in short supply, but things have a way of working out so the buyer got his order.

Completing the logging operations on the Quinault in 1942, we purchased a stand of old-growth Sitka spruce on the north side of the Queets River about a half mile above the mouth of Salmon River. The property was owned by Polson Logging Co. and the Merrill & Ring Company.

The Mayr brothers and a Mack full of spruce, in 1943. I am on the ground. On the truck, from left, are Harry Fry, Marzell and Otto Kestner.

Looking at the timber and the prospect of bridging the Queets, we went over a three-span cable foot bridge installed by Jefferson County so people on the north side of the Queets could get to the outside world. There was a single-track dirt road across Tacoma Creek leading to a big bench where still stood pioneer era homestead houses and buildings.

The war production board suggested this timber be made available for the war effort, and our company was ready to open up the stand and get it to the sawmills. We purchased the timber in early June of 1943. By July 5, we had bridged the river with nine full-length spruce logs, each 115 feet long. We built piers in the river with logs lashed together to form cribs 40 feet long set parallel with flow of the channel. We laid the stringers on the cribs with tractors and arch in the water, lashed everything together and planked the deck with cross-ties and a running deck of three rows of four-by-twelve-inch spruce under each running wheel. The bridge stood up well.

In all those years, Marzell looked after the logging and road building, and it was Marzell's innovation and engineering that completed this project. It held standard log truck loads of spruce containing up to 9,000 board feet. The stringers between the bents or piers sagged ten inches when the load went across, and then sprang right up again.

At this location we built our first portable logging camp set on logs, including a cookhouse, dining room, three bunk houses and a filing shack.

Among the timber cutters working with us at the time were brothers Larel and Norman Carlyle and Ed Fishel from Neilton. It was very difficult to find men that could cut timber of this size, but they were superb, the best. All cutting was still done by hand, and some of the large spruce requiring a ten-foot-long falling saw. Spruce are often swell-butted, and it either required a lot of spring-boarding or cutting the tree somewhere in the swell.

We always prided ourselves in producing a well-manufactured log, properly marked, with adequate trim and knots cut flush to the body of the log. These spruce made five 40-foot logs and the top three logs had horrendous limbs eight inches in diameter that had to be cut off flush. A spruce limb is so damn tough there is no way to remove the knot except by cutting it off with a limbing axe.

Besides setting exemplary production records our group was tops in payroll deduction for war bond drive.

Our tenth year in the logging business was 1943. At the time we were building an excellent banking relationship with National Bank of Commerce at Aberdeen, even though prime was four percent and they were not timid about charging us eight. To the bankers we were uneducated, uncultured, hard-working people and, as long as we made money, good for the bank.

After mid-September, we began to worry, "What if we got an early freshet and it took the bridge out to the Pacific Ocean?" By the middle of November, there had been a few scares. The river would come up a foot or two but not enough to move the bridge. We worked like hell and, by Thanksgiving all the logs were removed. We dismantled the bridge, cut the stringers into logs, removed any spikes and sent them to the reload.

During this same period we had a crew cat-logging hemlock on a tract near Tulips (or Wilderness), property left over from Gus Carlson Logging Co. It had been two-thirds logged by Art Shelgren in the late 1930s. Otis Hallin of Grays Harbor Pulp & Paper needed the logs badly. We moved in and logged while the season was dry. It was small hemlock, about 90 years old, and real good for pulp chips.

The company was becoming active on State timber sales in the Queets Corridor. When the Olympic National Park was formed, a mile-wide corridor on each side of the river from the National Forest down to the reservation boundary was condemned. These were divided into the inner corridor and the outer corridor, a half a mile wide each.

On the outer corridor the Federal government did not buy the timber from the State, but gave the State 20 years to remove it. There must have been 3,000 acres in the outer corridor averaging about 40,000 board feet per acre of a minimum size economical to log and truck that distance, 40-foot logs with an 18 inch top.

At a sealed-bid auction at the Jefferson County Courthouse in November, 1943, there were five high bidders under the war price board ceiling. The other four bidders agreed Mayr Brothers should be awarded the sale located on approximately 400 acres in Section 12, Township 24 N., Range 11 W. During 1943-'44, we cat logged 18 million board feet of beautiful hemlock and Pacific Silver fir, some fine quality old-growth spruce and Douglas fir, which was shipped through the Polson reload and rail line and processed by local mills for the war effort.

During the summer of 1943, Jennie and I bought the Charles Fry home in Cosmopolis at a prime location with a quarter of the block, at 318 East Third Street. The home was run down. We changed the stairway from the front of the house so it came down between our bedroom and the kitchen. We had Mr. Alwood build a new two-car garage with storage room and a small loft with a car port on the face side. It was a good place to store my row boat.

Marzell and I each drew a salary of $150.00 per month during those years. Jennie and I stayed close to home during the war years. We were able to save money on that salary.

When we first logged on the Queets, nothing much had been logged from the Bureau of Indian Affairs Quinault unit, east of Highway 101 clear up to Forks. There had been some logging along county roads on the Queets, Clearwater, Hoh and Bogachiel Rivers, but no roads were built into the foothills. The Civilian Conservation Corps had built a one-lane road and bridged the Clearwater just below Coppermine Bottom. The bridge was of box truss construction built with treated timber but it was not suitable for log loads.

In 1943, the Queets valley still had pioneer settlers living on their places. George and Ted Anderson lived in their original homestead house, but it was sinking into the ground. All the foundation was decayed. Mr. and Mrs. Harry Kitridge lived on a short side road below the Salmon River bridge and Mr. and Mrs. Ransom Higley lived about a mile above Salmon River. About five miles up, Pete Sutton and Carl Bush had a semi-portable camp and were logging for Lud Esses.

Logging and loading old-growth spruce and hemlock with Cletracs and carco arches, in the early 1940s.

Up above Matheny Creek, Mr. Gwin had a place on the river and lived out his years there. Then came Kelly's Ranch, a sort of lodge with cabins for rent to guests. They arranged horse caravans into the upper valley to Oscar Smith's place on Tshletshy Creek which could only be reached on foot or horseback and then only when Sams River and the Queets River were fordable.

In the winter of 1943-'44, we moved our logging camp up to Section 12 and set it up on the plateau above the Queets River.

Joe Nieradzik was our first cook at that location. He was a long-time cook with Polson Logging Co. and a good one. One evening I went to the cookhouse after dinner and thought I complimented Mr. Nieradzik on his excellent beef sandwiches. We had obtained some Argentine canned beef in two-pound tins and it sure beat Spam.

Next morning he came to me and said, "I am leaving on the next truck." When a cook decides to quit, that's it, no questions, no argument. They are very temperamental, just like a saw filer.

Our portable logging camp on the Upper Queets.

I found a young fellow who had just married Jack Brim's daughter at Humptulips and had just enough logging camp cooking experience. His name was James Trantham. He came up from Arkansas. He was meticulous and very clean about his person. He had a hard time getting along with "bull cooks," but Jimmie stayed with us until we moved our camp out of the upper Queets. The bull cook was the assistant cook and camp flunkie. He waited tables and washed dishes and kept the stoves going.

During the summer of 1944 we produced over two million board feet of log scale per month. This was all tractor logging, and we loaded with a loading machine converted to diesel using nine-and-a-quarter by 10-inch Tacoma-built steam donkey drums. The haulback had to go through a fairlead because the drums did not line up with the loading tree. We used a spreader bar and two tongs. Pete Raich loaded for a long while, until he was called into the service in the fall of 1944, about the same time as Harry Fry was called in.

One of my jobs was scaling in the woods for the cutting crew, which still had some old-time hand-cutters. These men usually went to town every two or three months, and hung one on. Sometimes, I would get them out of the Aberdeen or Montesano jail.

One fellow, John Buzee, a Lithuanian like George Stein, used to leave money with Mr. Stein on the days he went to town, as did others, so he would have enough left for a "sobering up bottle."

One time I got George Stein to round Buzee up and on my way to camp Monday morning he rode with me. He was bound I should stop at a butcher shop in Hoquiam so I went into the shop with him where salamis were hanging from a rack. Others, including women, were in the shop when he took one he had specially picked out and said, "I take this one. It's hard, just the way the womens like 'em." I just about sank through the floor.

By the time we got to camp two hours later, he was so sick he took his whiskey bottle and smashed it on the rocks in the parking area. Then he crawled into his bunkhouse and we never saw him for three days. He finally came out for breakfast looking wane and haggard and in a few days he was back in shape again, working every day until the next time.

John Buzee did not last very many years and I lost track of what finally happened to him. He stayed in camp on weekends and carved spoons out of vine maple. They were masterpieces. Jennie and I still

"Doctor" Doski, old-time timber cutter. His cold "prescriptions" always started with whiskey.

have one of them.

There were others, though they did not leave the same mark in memory. John Doski, nicknamed Dr. Doski for always recommending something for sick loggers, was of Russian extraction. A few that worked in the cutting crew during the war years were Oscar Natti, a Finn; Rudolph Schultz, a German; "down-the-hill" Conrup Wierup (His nickname came because he recommended falling down-hill to "save timber." Really, he hated to wedge.) and his partner Clifford Spark; and Mr. Ahlquist, a Norwegian.

The saw filer was a key to a good cutting crew. The stronger the man the longer he wanted his rakers, until it would pull out a long "macaroni" shaving. A smaller man played out pulling that much wood out of the cut at each stroke, so often the saw filer got run off.

One of the most powerful set of fallers we had during the war years was Norm Carlyle and Ed Fishel. They were in their 20s and I saw Norm Carlyle take out a double chip on the under cut. A normal chip was four inches. He would work in a neat notch on the stump and them come down from the top and break out a chunk of wood seven inches long.

Two other good buckers were Vic Dasher from Clearwater and Edmund Ness, a Norwegian and part-time fisherman. I remember when he came looking for a job. He said, "I am the best God-damn bucker in all of Grays Harbor." He and Vic regularly bucked 500,000 feet each of logs every month.

In June, 1944, a one-armed man came out to camp. His name was Clarence Baker and he lost his arm in a felling mishap up in Skagit County. I saw he was a qualified man for the cutting crew. I asked, "Can you handle bull bucking job, including woods scaling and keeping time?"

"Yes," he said, "I certainly can." He got the job and worked for us for twenty years and always made good money but put it into a holly farm at Humptulips City and the terrible cold in 1949-'50 killed most of his trees.

Marzell and I both looked after the trucking, which was a nightmare during the war years. Two million board feet of logs per month took quite a few trucks. We could get three, sometimes four loads to Polson Logging Co. reload per day for each truck. We had poor tires and poorer metal in rear ends and transmissions. Also, we hired quite a few common carrier trucks. Asa Fishel with his truck worked for us during that period.

Finally, in the fall of 1944, the International dealer in Aberdeen had three brand-new KR11 single-axle logging trucks available. We went to work with Fred Norman, our representative in Congress. He was from Raymond. He got an A-1 priority for us and we got the trucks in a matter of days. Those trucks, compared to our old Fords, performed very well for about 25,000 miles, but then the bearings were shot in the transmissions and rear end.

In January 1944 we bought our first section of timber land, Section 26, Township 20 N., Range 9 W., on the west branch of the Wishkah.

There was no road within two miles except the old Greenwood grade, which went through the section. It was river logged by Al Coates between 1898 and 1900 and Blackwell's dam was on the property. Later, State Fish and Game blew the dam out.

I remember going to our beloved banker, who was then Clyde Pitchford. I asked for $4,500, the price owner Mel Moe wanted. Mr. Pitchford asked, "How much old-growth fir and spruce and cedar is on the section?"

I said none. It had been logged 40 years before but it had a fine stand of residual hemlock and some fine 40-year-old second-growth fir and hemlock.

He hemmed and hawed but finally agreed. Later that year we bought 140 acres of old-growth fir and hemlock located on the upper Queets from our good friend George Kellogg. It had three million board feet of fir and spruce and five million board feet of hemlock at a price of $16,000. Timber land in those days was priced at its value at time of purchase and could be harvested at a profit. Usually, residual timber was left in long corners or steep draws. Title to the land was extra gravy. The logging by itself was a reasonably profitable operation.

We completed most of the logging on Section 12, and that winter we opened up Section 21 in the same Township with roads. Timber on half of this section was purchased by Grays Harbor Veneer Corp. owned by A. J. DeLateur, his brother-in-law Lou Cole and Puget Sound National Bank. Our relationship with the Veneer Corp. worked well from 1944 through 1964 when they sold out to Anderson Middleton. Grays Harbor Veneer and our company complemented each other during the war years and post-war years. A mill without a productive logging company would not have been able to keep up with their log needs.

Grays Harbor Veneer, having good banking connections with Puget Sound National Bank, was able to make funds available to us at prime rate for purchase of timber and timber land. In 1945, before the war ended, we jointly purchased the timber on three-quarters of Section 12-21-9 from Milwaukee Land Co., a subsidiary of Milwaukee Railroad. The purchase contained at least 25 million board feet. We purchased the remaining quarter section from Clive Abel along with an old 11 by 13 Tacoma steam yarder and two steam duplexes which we scrapped out.

We were logging Section 21 in August, 1945, when news came out by truck driver that the Japanese had surrendered. We kept right on working. After all, our many friends in the service certainly couldn't walk away just because the conflict was over.

At the time we had two people on medical discharge. Horace C.

Alwood contracted malaria in the South Pacific, and John Rycz had his leg crushed in a gun turret aboard a cruiser.

The war ended in August, 1945, but the effects of the war lasted many more years. The young men coming back from the service wanted jobs, but they did not have any logging experience. In one year we lost two young men in logging mishaps. Young Curtis suffered crushed vertebrae in the lower back and lived about a month in the hospital. No one ever knew how the other, Wonsewicz, got hurt. There were no witnesses. He died very quickly. Memory of loss of life by a person working for you is something a person never forgets, even 40 years later.

Marzell and I had worked together twelve years in 1945, and we saw the logging company taking shape as a long-time business relationship.

Jennie and I could not have children, so through Catholic Children Services we inquired about adoption possibilities and procedures. We adopted our first baby, Catherine, in September, 1945. She was born in June, 1945, and needed a family that could give her a home. The first adopted was a strong, beautiful baby, and many more followed. Jennie just loved babies and I thought it was a very good way to go.

By 1945, we desperately needed a shop so we could work on our trucks out of the rain. The building we rented belonged to Lennon & Wyman who had used it as a blacksmith and welding shop. Grays Harbor Railway and Light had used it to house their electric street cars during the 1920s. It was located in about the 400 block of East Market, across from Aberdeen Cabinet Shop. It was a lousy place. You could throw a cat through the spaces between the boards.

We were there not two years when we moved to a building on West Simpson at Hoquiam that belonged to Leo Kosenski. It was a little better, and, at least, closer to the work.

In September of 1945 we moved our logging camp down from the upper Queets and set it up on the old Polson grade about a mile east of Donkey Creek. We were getting ready to log Section 12-21-9. Our road traverse was originally the Humptulips-Wynooche truck trail. The U.S. Army had a camp on Stevens Creek just before the war ended for the purpose of pushing the road through. Unfortunately for us they pulled out before the road got through the tough rock. We encountered our first hard rock and had not much knowledge on how to drill or blast it. By constant pounding and blasting we worked our way around a rock point.

Now of course there are roads throughout that whole country and clear up to the park boundary in some places.

About that time Ralph Blaine, just fresh out of the Coast Guard, was our bookkeeper, the first after myself. When I look at some of the old

Frozen assets. Our first high-lead machine, the 11 by 13 Tacoma steam donkey on the E. Fork of the Humptulips in January of 1946.

books I kept, I see I didn't even give the year except on the outside cover. You can still read 1942-'43-'44. I guess I thought a person could figure that out O.K., there still being 12 months in the year.

We had Western Steel retube the boiler and rehabilitate the bearings on the old 11 by 13 Tacoma steam donkey. We put lines on her and hauled it up to the setting in Section 12. We had no climber and not sense enough to find a good one and ask him to help us rig up on a weekend. We raised a five-foot butt diameter old-growth fir, 100 feet tall and after about a week we got the blocks up and the heel boom in place, ready to go. Marzell did the climbing. He must have been scared. I know I would have been petrified.

Like on all the old ground-lead machines, the haulback drum sat on the left side and had to go through a fairlead on the head block to spool correctly. The first time we strung the haulback out we grabbed about five or six roads. We were going to drop in after each road was finished, but you know what happened! The lines would not clear and sawed into the logs until the engine could hardly move the haulback, let alone drag the main line and butt rigging out. And slack... there was no way to get slack. We relaid the haulback so it would clear logs and trees, and then it worked pretty good.

That winter of 1945-'46 we earned our diploma in high-lead logging. When I look back, I think it is a wonder we didn't have some serious accidents during those first high-lead years, but we didn't.

We used to hear the Polson Logging Co. steam whistles coming from across the east fork when the wind was right.

Marzell always went up the hill early in the morning and fired up the Moxley so we would have steam up when the crew got there. He hated that old yarder with a passion. There was no water above us and no cleared area, all was timber. We had to haul water up that long hill to keep the water tank full. Bob Hiner was our first engineer and later he taught John Rycz.

After a few months, spring came and we started getting a few experienced high lead loggers. Otis Fairchild and Oscar Brulevald were good rigging men. They knew how to lay out roads. Later, we also had big John Johnson, the ex-bootlegger and whore house operator. He knew how to lay out the settings and hated running up and down the hill fighting hang-ups.

In the winter of 1946-'47 we built more roads so we could reach all of the section. Those were all rock roads. There was plenty of good material for ballast available.

Frank Hannick was hooktender when we were logging the back side over toward the east fork. There was quite a bit of large old-growth fir, heavy logs on steep ground. The boiler on the yarder would lose pressure and drop from 220 PSI (pounds per square inch) down to 85 PSI and the log would stop.

Frank worked a system with John, the engineer. "When you run out of steam, 'dog' her, give a weak toot, and we will answer, 'stop everything.' We will tie the log to a stump and let you know. Then you can let the log settle into the strap, let your friction off, and then go back and stoke your fire."

In a few minutes John would have boiler pressure up to where it was about ready to pop off. He would give a sharp toot and the crew would answer, John would slam on the friction, open the throttle, move the log ahead a few feet. The crew would blow "stop" and John would hold it with the friction, the crew would unhook the strap and give the whistle to go, and, with full boiler pressure, the log would go up the hill to the landing.

We purchased a brand-new Washington Model 203 diesel yarder in the fall of 1946. It was powered by an eight-cylinder Buda diesel engine. The engine was a 1250 RPM. At that slow RPM the torque converter would not transfer the power to the gears and the log just would not move if it was a big fir.

Once, I was out in the rigging trying to figure out why the machine would not pull. We stood on the top end of the tree on which we had hooked on the butt log. You could hear the motor whining away but nothing moved and I asked the cutter, Harold Hunt, "Is this log bucked clean or does it have a Russian coupling left holding it?"

"No," Hunt said, "it is bucked clean."

The upshot was that we bought a Cummins NH200 that turned up 2300 RPM and that brought the logs out in good shape.

We used a heel boom for loading throughout the '40s and early '50s. Some loggers were using a new system called shovel loading, using a used construction shovel. It was usually a one-and-a-half yard capacity machine, with a heel boom installed to hang the tong from, using the main drum of the shovel to activate the tong.

In 1947, there was a boomman strike that lasted all summer so Baker and I felled timber "by hand." With only one hand and arm, he could swing that double-bitted felling axe just as good as I could with two arms. When we sawed the back up he would take the opposite side of the tree for convenience of sawing. That summer we laid down most of the timber that needed to be cold decked. We later swung it with 2,000 feet of two-inch fixed sky line using the Washington 203 as main yarder.

Before we finished Section 12 and the U.S. Forest Service sale we bought on Donkey Creek Ridge, we had some more interesting side lights.

The cook, George Perry, got fed up and quit because the water line froze up, so I got another cook. In a few days I caught him peddling beer to the crew. I canned him right then and there. It was a little nip and tuck finding another cook, but we made it.

Another time the cook woke us in the middle of the night, to tell us the camp was on fire. It was the dry shack burning. Somebody hung their underwear too close to the stove. We hooked on the end straps of the shack and pulled it away from the main camp and saved the camp. The dry shack and a lot of cork shoes and rain gear were a complete loss. Our insurance paid for the loss and we replaced the crew's clothes. In a few days, everything was going again.

About that time we got a good builder, "Frenchy" Lampertz, from Cosmopolis, to build some really nice bunkhouses and a new cookhouse and dining room. They were lined inside with plywood and painted white. They had electric light and all the good stuff, including a nice diesel heater stove. The buildings were 12 feet wide, 32 feet long, with 7-foot ceilings, and usually had accommodations for five men.

The time we logged in Donkey Creek, from the end of 1945 to winter of 1947-'48, were good years for trout fishing in Donkey Creek.

Some of the deeper and larger pools were especially good for fly fishing. I caught some up to 14 inches. The west fork of the Humptulips had lots of sea run trout, usually after the first rains in the fall.

While we were on Donkey Creek, the Forest Service offered a large 35-million-board-foot timber sale. There were five bidders; Harbor Logging Co., Anderson Middleton's logging outfit; Saginaw Timber Co; Grays Harbor Veneer; Picco Logging Co; and Mayr Brothers Logging Co.

It was during the period of the Office of Price Administration ceiling, so there was a maximum ceiling bid and the bids were all ceiling price. The Forest supervisor, Carl Neal, made the decision as to who should get the sale; the bidder he felt could do the most to expand the availability of forest products for the country now at peace. He awarded the sale to Ed Picco. I think he made a good decision, though we and Picco tangled frequently at land and timber auctions.

In January, 1948, we completed logging in the Donkey Creek area, and moved our camp up to the Queets Valley and set up on an old homestead clearing on a bench above Matheny Creek, about a half mile south of the county road. We turned off about 500 feet above the old Matheny log bridge which was built by U. S. Bureau of Roads as part of a war effort timber supply access road. We set up camp on U. S. Park corridor land. Each time we moved camp we set up a better camp.

During February of 1948, Jennie and I adopted a beautiful three-month-old baby girl and named her Patricia, again from Catholic Children Services in Seattle. Catherine was two-and-a-half years at that time. We had made the Fry house in Cosmopolis very comfortable and convenient.

My sister Margaret resigned from the convent. She came to live with us for a while and then moved to Columbus, Ohio, to work at her profession as registered nurse. Later that year she met Richard House from Little Rock, Arkansas, and they were married in 1950 at Columbus.

Mayr Brothers moved back up to Kelly Ranch Road on the Queets to log about 30 million board feet on the north side of the river. Marzell engineered a bridge crossing the Queets River at King's Bottom. On the other side we hooked on to the old wagon road just above Tacoma Creek. This bridge was over 340 feet long, and there was another 100-foot bridge across a lagoon which filled with water when the river came up.

The bridge was built to stay in place during a rise of six feet of the river. The river came up seven feet and when you stood on the bridge you could hear the rocks and gravel rolling downstream from the force of the water. Unless you have seen the Queets boiling, you have no idea what a wild river it can be.

The bridge stayed in place the winter of 1948-'49 and the winter of 1949-'50, although several times we had to adjust the stringers on the upper side because the piers in the river sank from the terrific water pressure. When the water went down, Marzell and his crew put a full-length tree under each pier, after that it never sank any more during a freshet.

While Marzell was building bridges and roads on the north side, we were logging a piece below Pete Sutton's camp, and 40 acres of old-growth fir we purchased for our friend George Kellogg during the war. It was good-quality Douglas fir and all was sold to Anderson Middleton and cut into timbers for the Australian market.

Clarence Baker, our bull bucker, and I located an 80-acre timber claim that was not far from our new road system and cruised the piece and made value analysis. It had some nice old-growth spruce, a few Doug fir and the rest hemlock. One old fir snag was so huge and heavy we would end up leaving it on the landing. It was 12 feet in diameter on the big end.

Contacting the owners in Seattle, they came down to Aberdeen. They were Granddad Baer, Mr. Baer and young Baer, like Grandpa Bear, Daddy Bear and Baby Bear, so we called the property the Baer Claim. We closed the deal in C. W. Adams' office in 1948, paying $8,500 for about one-and-a-half million board feet.

Including the Baer claim and parts of three or four sections up river from the bridge, we had about 30 million board feet to log. My good friend George Kellogg, liquidator for the E. K. Wood Lumber Co., called and offered us 500 acres on the north side of the Clearwater River and a section of hemlock about a mile south of the junction of North Beach Road and the Dekay Road. That section was mostly 80-year-old hemlock which, with the high road development cost, was of no interest to us. We could only buy what either our beloved bank or Grays Harbor Veneer would finance for us.

We considered the piece on the Clearwater worthwhile. It had old-growth spruce, cedar and hemlock and it straddled Elk Creek and several ridges which made it strategic to the logging of adjacent timber. The very next year Rayonier moved Morrison Logging into their Clearwater holdings to salvage hemlock damaged by the hemlock looper, which was beginning to defoliate the hemlock in the Clearwater.

So they started a gigantic salvage operation and it came about they wanted our E. K. Wood purchase. We negotiated a trade for 450 acres of old-growth spruce, hemlock, White fir and a few Douglas fir located on the west fork of the Humptulips; good timber and an easy logging chance. One quarter-section cut out 10 million board feet.

We were trucking all of our logs to our own log dump located above the mouth of the Little Hoquiam River, Esses Dump and Log Boom. There we could do a better job of merchandising. At that time we were selling all our peelable hemlock, White fir and smaller spruce to Grays Harbor Veneer Co. Hemlock saw logs were sold to Bay City Lumber, Wagar Lumber Co., Anderson Middleton, American Mill Co. and the Hoquiam pulp mill.

Hemlock logs became very popular right after the war. The Canadians had not got in gear yet for the expanded market and there was shipping space available from Northwest ports to the east coast where the carpenters were used to green hemlock lumber.

Calfornia builders seldom used green hemlock lumber. The California market specified Douglas fir lumber if available and when Washington fir stands were pretty well cut out, they were able to get it out of Oregon and Northern California.

The market for logs deteriorated during the summer and fall of 1949. Finally, the demand got so bad no one was buying any saw logs. We were still able to sell peelable logs to Grays Harbor Veneer and spruce logs to Woodlawn Plywood. In order to keep going we put up about 15 rafts of saw logs and got permission from Rayonier to store them on part of their south bay storage area. When that was full, we decked logs in the woods along the road.

In November, terrible storms came one after another. In the meantime I tried to get Edgar Anderson, log buyer for Anderson Middleton, to buy a few rafts. After deliberation Edgar said, "I will try one raft. Send it over." On the day the tug brought the raft to their mill, Edgar and I were standing on their dock on the log slip side where we could see what the logs looked like. Mr. Anderson let off a stream of uncomplimentary cuss words and said, "I don't want that raft. You have high-graded all the best." I never said more than "Yes, Mr. Anderson,"and instructed the tug to take the raft back to the Rayonier tie up.

The storms continued on into December and between Christmas and New Year, it got cold and froze every night. After New Year it started to snow and around the 12th of January the thermometer fell to -8 Fahrenheit at our logging camp. Our two-inch galvanized water supply line froze from one end to the other.

The snow got up to three feet deep in the logging area. The thermometer slipped to between 18 and 10 degrees many a night. Aberdeen and Hoquiam had snow on the streets for over a month.

The water supply in Aberdeen was interrupted by a break in the line about five miles below the head works. At that time Aberdeen's 28-inch diameter wood-stave line was laid on top of the ground so when water stopped running, ice formed which hampered getting the water flowing again.

In Cosmopolis there also was a water problem. Cosmopolis was getting water from Mill Creek, as they had since early times. The water system worked like this: The water flowed by gravity from a sort of beaver dam about one-half mile above town, through a 20-inch wood-stave pipe to just across the 101 Highway right in town to a pump house. The pump was turned by a Washington one-cylinder E- step diesel engine and it ran day and night to keep up pressure in the distribution system. You could hear that old engine, especially at night. It sounded like, "ka tuck - ka tuck." When that engine stopped that January, all hell broke loose. People were melting snow and boiling the water.

The water system used to belong to Grays Harbor Commercial Co., but when they deeded their timber south of Cosmopolis to Neil Cooney, he also got the water system. When Neil Cooney sold his timber to St. Regis Paper Co., they became providers of water for the town.

Ralph Blaine and myself ran for City Council that February on a "get good safe water for Cosmopolis" plank. We got elected and went to work looking for better water supply. We were able to buy water from Aberdeen and they laid a 12-inch cast-iron main from their city limits across Mill Creek and hooked us in. Ever since then the city of Cosmopolis has had good water.

The Council, Lord Mayor and city attorney, were able to negotiate a deal letting St. Regis off the hook. The company was under pressure to get the system fixed. They had all the land and timber south of town so we told them to sell to the City, for $1.00, certain property on the hill which had some timber. The City got the property, sold timber and lots and got good water at a reasonable cost

That winter, logging came to a complete standstill for about four weeks, but the road crew was clearing snow on up above and there was a cook and small crew at camp. On a Monday I started for the logging camp and was only able to get about five miles up the Queets Valley road. There was just too much snow. Having three five-gallon metal cans of milk, I cached them in the woods off the road. I knew they wouldn't spoil for a week because of the cold, and probably would even freeze solid. The following week the road was open and I picked them up and brought them to camp.

One other time in the summer on that damn bumpy Kelly Ranch Road I had five watermelons on top of the pickup load and they rolled off on the road. Did the cook ever give me hell! I wasn't very popular around camp that week.

One day in January the phone rang. When I answered, a voice said, "This is Edgar Anderson speaking." We exchanged pleasantries and he said, "By the way, Werner, do you still have those hemlock rafts at Rayonier storage?"

"Yes, Mr. Anderson, we do."

"Well," he said, "bring them in now."

I said, "Mr. Anderson, there are quite a few logs lost from those terrible storms."

He said, "Bring them in."

We agreed on the original scale at a price of $28 for No. 2 and $26 for No. 3.

Logging is a hell of a game. I would never advise people to go into logging the way we did, owning the timber, doing the whole job from standing timber to the finished raft. Tide, wind, and the winter's mud, the summer dust and intricacies of the market all punish your soul and upset your plans. For one time good timing, there are nine times bad timing.

Jennie and Cathy (four years of age), stayed at the Matheny logging camp in the summer of 1949. Cathy still remembers. We were the only ones there except for the watchman, Mr. Ranta. I thought it would be something for Jennie to remember. Logging camps were getting fewer all the time, so much of the timber in other areas was cut over.

The largest stands of remaining timber were in the Olympic National Forest, which, before the Olympic National Park was formed, was one of the largest National Forests in the state and would have supported an economy similar to Eugene and Springfield, Oregon. Over 25 billion board feet of old-growth timber were removed from the economic well being of Grays Harbor with the forming of the Park. This timber stood on 300,000 acres of some of the best timber-growing land in the world. That number of acres of land would produce a sustainable harvest of 240 million board feet every year. Even today, after 50 years, people question the size of the Park. The upper valleys certainly would have been enough park.

In 1948, we took delivery of one of the largest log loading machines ever built, a Washington TL21 Track Loader. With it, we were able to load 40-foot cedar logs, containing 9,000 board feet of net scale which included the very large cedar logs common in the University block.

At the time of Statehood, the University of Washington received from the United States public domain a whole township of fine timber on very favorable terrain. We logged quite a few sections of this timber in the years between 1950 and 1965, and most of this timber was hauled to the Harbor. Many of these large cedar logs were delivered to E. C. Miller Lumber in South Aberdeen. We sold cedar logs to Millers through three generations.

Toward fall of 1948, we completed some of the logging in the Queets Corridor and started logging 100 acres of E. K. Wood timber on the upper head of Killea Creek, above Kelly's Ranch.

There were two Sustained Yield Units set up in the Olympic National Forest. One was the Simpson Cooperative unit. In that case, the company pledged their tree farm for 100 years and the U. S. Forest Service pledged the entire Wynooche-Skokomish drainage and together this vast acreage of timber and timberland guaranteed continued economic life to the communities of Shelton and McCleary.

Most of the Olympic foothills in the Salmon River, Matheny Creek, and Sams River upper drainages were National Forest land. The Quinault Ranger District was in charge of administration of these lands. In 1946, after hearings held in the Morck Hotel, these lands were set aside for exclusive processing in the Grays Harbor area as the Grays Harbor Federal Sustained Yield Unit Number One.

In Unit Number One, there was no contribution of timberland except the National Forest acreage. The acreage growing timber was less than half of the Simpson Unit, so the annual cut was much smaller.

The sawmill economy on Grays Harbor was shrinking. Wilson Brothers mill went out, Northwest Mill shut down, Donavon No. 2 closed and Eureka Mill burned down.

For Jennie, my brother and I, the 1940s were years of development and expansion. Toward 1946 we became high-lead loggers. Cat logging with Carco logging arches was history. We were logging with one high-lead side and a smaller triple drum, Carco winches on a Cletrac tractor and, once in a while, our little Skagit BU110 sled machine.

We pushed our road on the north side of the Queets clear up to Section 1 across the river from Kelly's Ranch, but there was no crossing the river at that point. When I look back, a person had to be either grossly dedicated or slightly dingy to tackle country so far out and with such poor roads.

Before we got through logging on the Queets Kelly Ranch Road one of our trucks broke the stringers on one of the bridges on the Kelly Ranch Road, the Salmon River bridge, resulting in all of us being locked in, and no way in or out. All the stringers were rotten and needed to be replaced, and they had to be Douglas fir for a main road.

I went down to Anderson Brothers' ranch and asked to use their telephone and called the Superintendent of Olympic National Park and the Queets corridor, which was not yet park status. I told him what happened.

He said, "My, my. What do we do now?"

I said, "Here is what we will do. We are cutting six old-growth Douglas fir trees, 44-inch butts with 30-inch tops, 110 feet long, located on U. S. Government land. Then, we'll haul them to the bridge site and rebuild the bridge." I hung up and we proceeded.

The Park Superintendent was discreet and never came around until the bridge was rebuilt. Then he came around and said "My, my. Don't you ever cut any more trees off U. S. Government land."

I said, "No sir. Absolutely. I won't do that again."

Marzell's bridge engineering paid off but nobody paid us for rebuilding the bridge which lasted many years. Besides the logs, many a happy steelhead fisherman came across that bridge.

Esses Logging Co. was logging and using the Kelly Ranch Road during those same years. They logged part of state sections 15 and 20 and also were successful bidders on two National Forest timber sales located on the north side of river. Carl Bush and Oscar Blechschmidt were their main foremen. Around 1946, Blechschmidt went logging for Olympic Hardwood in the Raymond area.

The old house we grew up in was built around 1885 and sat on wooden blocks and finally could no longer be held together. In 1947 Marzell and I had a designer draw up a new house for Mom and Dad. We found a builder named Bergstrom who built it for them on the same knoll right next to the old house. Our parents were in their 70s and we were proud to be able to do this for them. They moved into the new house in October of the same year.

In 1948, Jennie had some elective surgery at St. Joseph Hospital and there she met a young nurse named Frances Dionne. She had just come to the Harbor from New Brunswick, Canada, and she was lonely. Jennie said to her, "I have a brother-in-law and he isn't married."

Frances was the daughter of the Irishwoman Catherine Mary Ryan, a descendant of the potato famine Irish, many of which settled in Eastern Canada. Catherine Mary was born at Johnville, New Brunswick in 1888, the grandchild of 1840s Irish exodus.

She married Adolphus Charles Dionne or "De Yone." Out here in the far west the name Dionne is used. Mr. Dionne was born in Red

Marzell and Frances Mayr were married in St. Mary's Church in 1949.

Rapids (Isladwin), New Brunswick, in 1887. He was from a long-time French background.

We were unable to get further background on the Dionne family. The people living in the agriculture part of New Brunswick from the late 1700s or early 1800s were farmers working in the winter months in the forest, cutting timber or logging with their horses. They used sleds piled high with logs, traversing iced snow roads to the river bank from where, at spring breakup, the logs were driven down to the salt chuck for use as raw material for the sawmills.

By middle part of the 1800s, the old-growth White pine and spruce were all gone and logging was working on second growth, such as it was. No one reforested anything in those days so the trees growing back were more hardwoods and spruce, with some eastern hemlock and an occasional White pine. Of course, in that area nobody else had any old-growth left, so value and quality balanced out.

Anyway, in those days Marzell went to Sunday mass with Mother and Dad and Frances knew this so she went to see what Jennie was talking about.

Jennie was out of the hospital in a few days and she arranged an invitation for Frances and Marzell to a dinner party at a place called DeWitts Chicken Dinner Inn located on Cosmopolis hill. That evening, Cathy, then five years old, and I went to the apartment where Frances lived and brought her to our home in Cosmopolis.

Frances and Marzell met for the first time that evening. The following year, 1949, they were married in St. Mary's Church. Cathy was flower girl, Francis Hannick was ring bearer. Frances' uncle and aunt, the Murphy's from Satsop, attended her. Bridesmaid was Patricia Ferbache and Matron of Honor was Jennie.

Chapter 5: The Fifties

Cedar, Cedar, Cedar

THE EARLY WINTER MONTHS of 1950 ushered in some of the coldest weather we had seen in this country since 1928. We were getting ready to move our logging camp again, this time into a practically virgin area of the Olympic National Forest and the part of State forest set aside for the University of Washington.

During June of 1950, the United States became involved in the Korean conflict. North Korean soldiers marched across the 38th parallel into South Korea and caused the United Nations peace-keeping mission to commit military forces into the conflict. The United States and Canada bore the main manpower brunt of this conflict and, of the two, the United States by far committed the most manpower and equipment.

This conflict had an immediate effect on the economy of our country. The Forest Products Industry was again called to furnish materials for both a war effort and for the pent-up domestic needs.

During 1950, the U.S. Forest Service offered several large sales which required significant road and bridge construction. The first sale was the Humptulips sale of 40 million board feet located on the West Fork. This sale required approximately 10 miles of road and a concrete single lane bridge, named the Gorge Bridge, across the West Fork of the Humptulips River at the old splash dam location.

The timber auction conducted by the U.S. Forest Service was held in the old Aberdeen City Hall in May, 1950, opening at 10:00 a.m. The main contenders were Picco Logging Co. and Mayr Bros. Logging Co. After qualifying opening bids and deposits, the Forest Service auctioneer opened the sale for oral bidding. We both wanted the huge timber sale.

Bidding sometimes moves very slowly and requires considerable time to work out the numbers. Bidding continued without let-up until 8 p.m. of the same day. Marzell and I were sitting by ourselves and determined Picco wanted the sale so badly he might suffer apoplexy if he lost out and we figured we could find a sale later. We dropped out and the sale was awarded to Picco Logging Co. We waited our turn on the next sale, Killea No. 2, about 10 million board feet. We bid against Blagen Mill Co. and finally conceded the sale to them.

There was one more large Forest Service sale that year and that was Raft River No. 2, about 35 million board feet. There were five bidders on the opening board, Anderson Middleton, Wagar Lumber, Blagen Mill, G. H. Veneer Corp., and Mayr Brothers Logging Co. This was a completely new area of development, and the road and development costs were astronomical, but we wanted the sale. It had a big volume and good quality timber.

After opening bids and qualifying of bidders, the sale officer declared the auction open for oral bidding. After about four or five bids we were high bidders and we were awarded the timber sale. This gave us a backlog of timber for several years. With this volume we were able to negotiate with Blagen and obtained the Killea No. 2 timber sale, located in the same area.

Raft River No. 2 began three-quarters of a mile west of the Salmon River, and encompassed parts of half a township and some of the finest old-growth timber in the Olympic National Forest, 95 percent of which was hemlock and Pacific Silver fir. This contract required the building of 11 miles of new road and a single-lane concrete bridge 140 feet long over Salmon River.

After long negotiations with the State Land Commissioner's office we were able to lease two acres of land at the Salmon River crossing as a location for our logging camp.

Frank Hannick and I worked on mats with the Cletrac bulldozer and a new Lorain one-yard drag line pulling mud together, building toward the Salmon River. The huge hemlock stumps we split apart with stumping powder left holes that filled with water, making the entire right-of-way a quagmire. In one place we got the Cletrac at such a sharp angle the back main bearing ran out of oil. I don't remember how we ever got the machine level and out of there.

Spring came and the combination of less rain and the drying effect of the sun again made the soil workable. We got the main road rocked to Salmon River and moved our camp into place right on the edge of Salmon River, installed a septic tank and brought in a two-inch water line which flowed by gravity from the nearby hill.

It took more than a year to build the foundation for the permanent bridge because the sandstone was unstable for the footing and this required one additional pier and a 40-foot approach span. To clear the location site for the bridge, it was necessary to build a temporary bridge of log stringers and planks, and about one mile of extra road to get around the bluff.

Using a fixed skyline which Marzell rigged up we were able to slide the 60-inch by 100-foot fabricated main girders across the canyon and rest them on the far pier. Lamb Construction brought in steel men from Seattle for the riveting work. Once the main girders for both spans were in place, everything went smoothly. The false work was installed for the deck and bull rails, then reinforcing was installed and in a few weeks the deck was poured. The bridge opened June, 1951. It has since been replaced with a double lane bridge by the Forest Service.

On January 1, 1950, we incorporated Mayr Brothers Logging Co. Inc. We put our logging and trucking equipment assets into the new corporation as well as all of the timber contracts and log inventory. In those early years we were feeling very determined and dedicated to building a tree farm and we were aggressively purchasing timber land for it.

Tree farms were born in 1941 with the dedication of the Clemons tree farm by Weyerhaeuser Company. Rayonier purchased the Polson Logging Co. a few years later, which greatly added to their land base and Schafer Brothers, one of the oldest logging companies in the timber business, saw the need and began very aggressively to add to their holdings.

The industry was actively tree farming their holdings. Mainly it meant keeping wild fires under control by cleaning up the logging debris and control burning slash after logging and planting the land.

By early 1950 we held separately in our partnership 3,000 acres of some of the best timber land lying north of Aberdeen in the Wishkah Valley, most of which had been logged in the very early times. We planted our first plantation on one of the parcels bought from our friend, George Kellogg, the 40 acres on Matheny Creek.

The partnership never carried on any logging operations after our incorporation. We also at that time brought in an outside audit company, Price-Waterhouse of Seattle. They were recommended to us by Mr. Ned Bishop and our bank.

We thought we had a satisfactory relationship with Price-Waterhouse for many years, but as the years went by they kept sending different auditors to do our work. There was no continuity in audit personnel.

The log dump at North Hoquiam. We needed a "bulldozer with a snorkel."

Mr. Frank Yuskoff was our first full-time bookkeeper. He became ill and resigned and we were able to get Evar Carlson, who was with us until his retirement. He did a good job. Having been with Harbor Plywood Co. in Aberdeen, he had a good timber background.

All movement of logs in those days was on the river by raft, so we leased the Esses log dump on Highway 101, just north of the little Hoquiam River bridge. It was a hell of a poor place to dump and raft logs. There never was enough water or space to store logs and at low run-outs, we would get some of the damndest jack-pots under the brow log.

I would come down to the dump and Harry Jordan, a big 6-foot, 2-

incher, would say, "Werner, what the hell do you expect us to do? We don't have a bulldozer that has a snorkel so it can go under water and fish out the 'dead heads.'"

I would say, "Harry, you guys are doing a damn fine job...and remember, Harry, a busy boom man is a happy boom man." That would disarm him and his crew and bring out a smile.

At the suggestion of Marzell, we drove a long piling in the river opposite the unloading A-frame, surrounded it with a brace of piling, and lashed them all together to form a solid anchor for the tall piling. We then put a small two-drum gas donkey on the boom race opposite the unloader, ran the lines through a block on the tall piling, Harry Jordan hung a tong on the main line of this donkey, and this gave them power to break the log jams under the brow log and to lift dead heads out of the bottom of the river. That was better than getting a "bulldozer with a snorkel."

Jennie asked where the expressions, "Holy O Mackinaw" and "great coat" came from. Most of Grays Harbor boommen came from New Brunswick and Nova Scotia, Canada, between 1895 and World War I. They were used to driving on swift, fresh-water rivers, usually after spring breakup. They, or at least some of them, always had a big wool Mackinaw, or great coat, nearby on the river bank.

When a boomman fell into the icy water and was able to get out, he quickly went to the gravel bar, stripped off everything and put on this great coat and quickly controlled his body temperature. In the meantime, there of course being no electric clothes dryers nearby, someone stoked up the wood fire and hung his clothes and socks and shoes so they would dry. As a life saver, the great coat became a Holy O Mackinaw, used again and again.

In the early years, all mills were water oriented. No logs were delivered to mills by truck until the advent of cedar roof-shake mills, beginning around 1962. Within ten years all booming and rafting operations ceased and mills built after the middle 1960s were prepared to receive logs off dry land operations.

Early in 1950, Marzell and I decided to build a state-of-the-art maintenance shop with offices upstairs. We had acquired about four acres of tideland located between Highway 101 and the Hoquiam River. It fronted on the old highway and the Hoquiam City water main went right past the front door. We filled the property up to three feet above the all-time high tide, which was 16 feet in 1932-'33, the year water went over the docks on the Harbor.

Blagen Mill cut the lumber and huge timbers for the main building

The family together for Marzell's birthday in Mom and Dad's new house on the Wishkah. We all were home.

and Lamb Construction Co. built it. It is 100 feet by 100 feet with an 18-foot doorway clearance and an overhead traveling crane. To this day, there has never been a logging machine we couldn't put in the shop for repair, lock, boom and everything. We moved in in May of 1950. Now we had our own headquarters, right on Highway 101. We had "identity."

There is no venture or business with so many seemingly insurmountable odd problems as logging from stump to dump. We were happy that Dad and Mother were living and could see for themselves that we were accomplishing something. Dad was pleased, but Mom had reservations. Years before she said, "How can you run such a vast business with no training or education?" Mother forgot for the moment we were loggers. Dad knew loggers can do anything.

Maxwell Carlson was a well-educated man and became president of National Bank of Commerce, forerunner of Rainier Bank. He was the one banker I knew that knew the logging and timber business. He was the son of Gus Carlson, who had Carlson Logging Co. and was involved in Carlson & Callow. Dad worked for Gus in the woods when he first came from the old country.

By the early 1950s we were operating 10 logging trucks, hauling logs from the Quinault area to our log dump. Horace Alwood was put in charge of the shop and maintenance of equipment and servicing of the truck fleet. Ernie Hansen was night foreman.

Marzellinous "Max" Mayr, 1877 - 1952

In 1952 we suffered a blow. Dad passed away suddenly. I remember it like yesterday. I was working on Raft River No. 2 road right-of-way. It was in the early spring, Frank Hannick came up the grade and told me Dad was real sick. Frank was our brother-in-law. He and Hedwig were married the same year Jennie and I got married.

At Dad's funeral, he was honored by a Requiem Mass celebrated by his friend of many years, Father M. O'Donnell. He always called Dad Max. Dad was laid to rest at Fern Hill by Mr. Bielski, another friend.

Dad's death had an impact on Marzell and me. It seemed to mellow us and bring us closer together. Marzell was married now and had a fine wife, Frances. They lived in a brand new house Mr. Alwood built for them on the home place.

Mother was a strong person and continued to live in the new house Marzell and I had built for our parents a few years before.

We expanded our contract timber reserve base and began purchasing State timber sales on the University of Washington land in Township 24 North, Range 11 West, in Jefferson County. The first sale awarded us was every alternate 40-acre tract in Section 36. This required locating all four section corners and all four quarter corners, then running and chaining 16 miles of lines to properly locate the 40s under the contract.

State law at that time only allowed timber on State land to be sold by legal subdivision. Later, the Legislature passed laws which allowed the state to sell timber within their ownership by terrain and drainage description, the cutting area being marked out in advance of the sale by the State Department of Natural Resources.

In the 1950s our main log customers were Grays Harbor Veneer Corp., which bought 90 percent of our peelable hemlock and White fir; Anderson Middleton Sawmill, which bought our saw logs; and E. C. Miller Cedar Lumber Co, which we sold our cedar production to.

E. C. Miller specialized in old-growth cedar for many years. Mr. E. C. Miller ran the mill until his son-in-law, Bob Ingram, returned from service in the U. S. Navy. He was an officer aboard ship. He ran the business for many years and was widely known throughout the Northwest lumber industry. Later his son, Bob Ingram, Jr., managed the operation.

E. C. Miller Cedar Co. was our main cedar customers through three generations. Finally the mill ran out of resources when the old-growth cedar went to the cedar roof shake market which took off in the middle 1960s.

We also sold hemlock logs to Blagen Mill continuously. Blagen Mill management was under E. H. Macintosh and Len Skalley. Wes Fish was the money behind the mill. He married a Blagen. After World War II, the mill just sat there and I think he just provided some operating capital and borrowed on inventory and receivables.

The mill was always a well-managed operation and added payroll to the City of Hoquiam. They continued with the manufacture of dry hemlock, mainly two-by-fours in the longer lengths for the Midwest.

Wagar Lumber in Junction City were also purchasers of logs from us. They were harder to sell by water because they were not directly on the river. They had to feed logs from the river through a tide gate at high water into a pond from which the log slip entered the mill. They were some of the first people to use semi-dry land delivery. They were able to inventory logs on dry land and chuck them into the pond, using an old shovel to unload trucks, deck logs, and put them in the pond.

The Wagar mill was a log gang mill and had a 24-inch maximum-diameter limit. The mill got good recovery using thin saws. Fred Hulbert, Jr. was the person running the operation and was pretty much directed by Fred Hulbert, Sr., his father. They did well during the 1950s and later sold out to a newly-formed company, Western Lumber.

During the early part of the 1950s, Marzell and Frances, who were married by Father O'Donnell at a Nuptial High Mass, had a daughter Mary, born December 15, 1950, and a son Thomas, born March 15, 1952. Just a few months after Thomas was born, Frances was admitted to St. Joseph Hospital with severe back pain. A few days later she underwent surgery for a ruptured disk, was home a few days, then back in the hospital with extremely severe back pain. Frances never recovered and passed away in 1953 at the age of 31 of a post-surgery infection. We were all in complete shock.

Jennie and I did what we could. Mother was close by and helped some. Marzell, with two small children, suffered a great deal from this tragedy. Jennie was able to get her aunt, Jennie Ohlinder, to come and take care of the children, Tom and Mary, for quite some time. She was a strong woman, meticulously clean and orderly and this was a God-send for Marzell. In a few years he met a nurse named Snevah Havens who had been an officer in the Army. They fell in love and married September 9, 1954, and she was able to make a home for Marzell and his children.

When Patricia was five years old, Jennie and I decided we wanted a boy. We received Mike from Catholic Children Services in November, 1953. Mike was a big boy with fair complexion and blue eyes with blonde hair like his mother. He fit right into our family life and gave us much satisfaction and joy. He was our first son. Our home on 3rd Street in Cosmopolis had a large fenced yard and it was an ideal environment for children.

In July, 1953, a truce was declared in the Korean War. The boundary between North and South Korea was set at the 38th parallel. Our nation continued heavy involvement in Korea, having a large contingency of U.S. troop presence in South Korea.

The presidential election of 1952 was Dwight Eisenhower vs. Adlai Stevenson. Eisenhower won with 442 electoral votes and also enjoyed a Republican Congress. One of the pieces of legislation passed by Congress and approved by President Eisenhower was the Enabling Act. This act had a profound effect on the states of Washington and Oregon.

The Act directed the Bureau of Indian Affairs (B.I.A.) to permit Indian tribes and Indian owners of trust allotments to begin to take their place with the rest of U.S. citizens, allowing them to realize the monetary value of their heritage allowed for by the various treaties negotiated during the 19th Century.

In Oregon, one of the largest reservations, the Kalamath, was located just east of the Cascade Range in the southern part of the state. It was divided into three parts for the purpose of practical marketing, allowing for reasonable blocks that could be sold.

Kalamath being a tribal reservation, the tribal members would receive a large amount of cash which would give each the value of his or her estate with which they would start their lives with the rest of United States citizens. Part one of the Kalamath reservation was bought by the U.S. Forest Service and annexed to the Kalamath National Forest. Part two, approximately 80,000 acres, was purchased by one of the large, long-time paper mills located in the Northwest. Part three was never sold.

The Enabling Act had a direct effect on our company because we enjoyed close proximity with the Quinault Indian Reservation, an allotted reservation of some 230,000 acres with a very small acreage of tribal land. The rest of the tribal timber had been sold years before (the Milwaukee Unit).

The Quinault reservation was divided into several operating areas by the B.I.A. South of the Quinault were the Cook Creek Unit and the South Quinault Unit. North of the river was the North Quinault Unit, which was purchased by Polson Logging Co. on a long-term 25-year logging agreement. The next unit to the west was the Taholah Unit, purchased at competitive auction by the Aloha Lumber Co. and completed by our company in 1976. Immediately to the east was the Crane Creek Unit, purchased by Rayonier at competitive bid in 1947 and completed in 1983. The remaining areas to the north reservation boundary, the Queets Unit, including the Raft River drainage, were partially sold by allotment by three methods.

In the event the B.I.A. judged the allotee was competent and well able to manage their own affairs financially and otherwise and were of legal age, they could arrange the sale of their property directly and the U.S. would give them a fee patent title. In case the allotee wanted the Bureau to arrange a sale of their property it would be advertised and offered for sale at public auction usually held in Everett, Washington, at the B.I.A. office for the State of Washington. In the case an allotee was not competent to either sell their allotment or manage their affairs, the B.I.A. in the case of need, would issue a fee patent and sell the allotment for the allotee and hold the money in trust drawing interest. This also was the case with minor allotee owners. Trust property could not be transferred to a non-Indian person or company.

The Quinault Reservation contained some of the finest stands of old-growth Western Red cedar ever found in any part of the world. During the 1950s our company was a large producer of Western Red cedar logs, so it was logical and good business sense to appraise and become involved in the allotments located on the Queets Unit of Quinault Indian Reservation.

It was part of my responsibility to locate areas of good quality timber with owners deemed competent by the Bureau of Indian Affairs. Upon their ability to obtain a fee patent from the United States of America, they, armed with appraised value and we, with our own appraisal, could negotiate directly and we could obtain a merchantable, insurable title. We also kept abreast of the monthly timber sales held at the Everett office of the B.I.A. There we would bid for, and, if high bidder, pay the bid price and obtain a merchantable title from the United States of America in fee simple.

Beginning in the early 1950s, we commenced logging our purchases as we developed the road system, logging one high-lead side and continuing on that basis.

Our company harvested about 80 million board feet of timber off the Queets Unit, beginning in 1954. Seventy percent of the timber on the Queets Unit was old-growth Red cedar. About this time, the market for Red cedar logs expanded with the advent of hand-split sawn 24-inch-long roof shakes for housing construction.

This rough split look made an attractive roof and if the roof was well cared for it would last 30 years, five years longer than a sawn shingle roof. In the 1950s there must have been 40 or 50 roof-shake mills between Oakville and Forks.

Pressure from the cedar roof-shake market caused change in value needed for No. 1 and No. 2 old-growth cedar logs. The large market for No. 1 and No. 2 cedar logs for shakes had a tendency to skew the grading of No. 1 logs. Shake mills could use a cedar log that had defects like weather check, splits and bark seams. By sawing two feet long and splitting down to two by six inches, they recovered enough clear wood to do real well on a log which was graded No. 2 per cedar rule, but yielded more clear wood for a shake mill than a No. 1 cedar log would.

In the early 1950s, I became a member of the Grays Harbor Log Scaling and Grading Bureau and a few years later became President of the Bureau. Six scaling bureaus, from Puget Sound Bureau down to the Northern California Bureau, belonged to the Pacific Northwest Rules Group. This group met twice a year, each time in a different part of their membership area.

In order to stabilize the grading rules for the timber and milling industry on the west side of the Cascades, the member bureaus sent their managers and Chief Check Scalers to seminars twice a year and conducted scaling schools. Information from these schools and mill cut-out studies did much at the twice-a-year meeting of the rules group to build integrity and respect for bureau work.

A positive result was that most public agencies selling timber at competitive bid, including the Bureau of Indian Affairs, State of Washington Department of Natural Resources and the U.S. Forest Service, accepted Bureau scale as the basis for cut-out grade and volume for their timber sales.

Some land owners did not use the Bureau scaling service, but their own scalers used Bureau rules and Bureau check scalers. The almost total acceptance of these scaling methods figured prominently in the valuation of billions and billions of feet of timber in the Northwest.

In the 1950s we widened the scope of our logging operation and increased our land purchases, each time with borrowed money, but timber values kept rising. In the 1930s hemlock was $1/M for a large tract. By 1955, hemlock sold for $10 to $12/M, an increase in market value of ten or twelve times. If interest was eight percent per year, compounded annually, the timber still cost only $3.50/M. Timber tax was ten to 15 cents per M per year, so the timber, after all holding costs, doubled in real value in 15 years. That made timber and timberland a terrific investment, especially if buyers were very selective.

To finance such a program and keep paying off our lenders, we needed to cut and market a part of our timber purchases sufficient to pay off the lender and secure clear title and do it at a profit. With good planning, we ended up having some merchantable timber remaining on the land and replanting the area logged.

We logged part of our lower Wishkah timber in 1952, along with timber off the west half of Section 29-18-9, which we purchased from our good friends Maxwell and Lawrence Carlson. Their father, owner of Carlson Logging, logged out the old-growth fir, spruce and cedar in 1905 but kept title to the land.

The hemlock off this property was sorted into peelers and saw logs. The peelers went to our regular customer, and the saw logs were hauled to Cosmopolis and loaded aboard rail car at a steam-powered reload and shipped to Rayonier pulp mill, located in Shelton, Washington.

Year by year the operation grew and became more complex. We never made much money, but we were building a timber portfolio, which had the blessing of the lenders and the forest products industry. Marzell did his part looking after road building and the logging. I could never do a good job operating yarders, loaders, bulldozers or road graders so I did what I thought was best for the company.

I always covered all details of the operations. I loved to work setting chokers or blasting stumps on the road right-of-way, laying out roads with Clarence Baker or whoever else I could find that could be spared.

Our work was brutally demanding. Sometimes we would argue, although we never got into a fight. Many years later we could look back and put our thumb exactly where and in which year we should have sold out, but like sailing a ship at sea, after the storm it became quite simple to see when the cargo should have been lightened.

In 1954, we purchased several quarter-sections of old-growth timber in western Jefferson county from Mr. & Mrs. W. E. Boeing of Seattle. Later we were able to square up the holding by trading with Rayonier and Peninsula Plywood, which gave us 1,000 acres containing about 45 million board feet on Kalaloch Creek. Negotiations for the larger blocks were often through a broker and was not as interesting as it was when we dealt directly with the owners. In the late 1950s we developed the tract and logged the timber along with our other operations. After logging we had the entire area hand-planted with Douglas fir seedlings.

A stockholder of the Grays Harbor Veneer Corp. named Andy Bator had a quarter-section on the ridge north of the Clearwater River. At the time, we had joined with Rayonier scouting property in the same area. They sent Dan Williams and George Lonngren, and the three of us had to wade across the Clearwater River just about at the end of the county road. It was late March, still pretty cold. I took my caulk shoes and socks off, and carried them on my shoulder with an axe in my left hand. The river was not swift, and, as I remember, about 14 inches deep. All of a sudden a big spring salmon swam up and went right between my feet, and me with no net! All I could think, "swat him with the axe." I didn't, because if I hit my leg I would have been in trouble, so that one got away.

While we are on fish stories, years before, about 1940, we were logging some cedar just above Lake Quinault on the north side. Our road went up a sometimes-dry stream bed. That particular morning, Gene Carlson, Marzell and I were going up the creek-bed road to work. It had rained pretty heavy during the night, so much that water was running down the tire tracks, and lo and behold, a nice four-pound Quinault salmon was slithering up one of the tracks. We had fish at camp that night.

In the 1950s, the State of Washington, under newly-elected Land Commissioner Bert Cole, began to develop the 95,000-acre Sustained Yield Forest Number One lying in western Clallam and Jefferson counties. Under Mr. Cole, the Department of Natural Resources was set up and various headquarter districts were organized, of which the Olympic District, located at Forks, was the largest. State timber sales no longer were held at the courthouse steps of the respective county seat of the county in which the timber was located, but at district headquarters.

This change in the law was a good and practical move. State timber sales were patterned after U.S. Forest Service timber sale methods, but in some respects the State Forest timber base differed from any other.

Cedar grew very large in the University Block

The State holdings were on the west slope of the Olympics, where the deep clay soils grew thick stands of tall hemlock. In November of 1921, a violent storm hit the north Washington coast and blew down thousands of acres of this hemlock old-growth.

George Anderson, pioneer Queets rancher, told me all paths, pack trails and road were completely crisscrossed with huge trees and you could walk long distances and never get on the ground, always walking on fallen logs. Very little of this blow-down was ever salvaged. There were no roads, and even had there been roads, there was still no market for such a huge volume of hemlock.

Nature took over and in a matter of ten years the under-story saplings and seedlings covered the fallen timber like a thick green carpet. As a result, instead of a single A-class old-growth stand, the State had a good percentage of thrifty second-growth hemlock already pole size when the market for these types of trees finally made the State forest profitable to log. By 1990 the blow-down had the same volume of saw timber ready to cut as the original stand.

Our company logged millions of feet off the University of Washington Township 24 North, 11 West. In most of this area there was not as much blow-down because much of the land was in the gravel-based Queets River basin.

In 1955 Jennie and I decided we would take a long automobile trip to Little Rock to visit Margaret, who was working as a nurse in a children's hospital. She was married to Richard House, whom she met in Columbus, Ohio. Richard suffered from lung infection and had contracted tuberculosis which he never was able, even with the latest medical attention, to arrest and bring under control.

It was in August when we got over the Rocky Mountains and stayed at Lamar, Colorado. We knew by the heat and humidity that we were on the Great Plains. The further east we traveled, the hotter the temperature.

In Oklahoma City, the thermometer read 95 degrees, Fahrenheit. We crossed the Arkansas River on a very narrow bridge. It seemed just wide enough for one car. We crossed into Arkansas at Fort Smith in thunder and lightning and warm rain. Driving with the windows rolled up was stifling, but if the windows were even partly down, bugs came in the car, and Jennie and the girls, Cathy, seven; and Patricia, four; and little Mike, 18 months; would set up a howl. They were terrified of the insects.

The same day we crossed the Arkansas River, we got to Little Rock. We stayed in a motel. Jennie and the children did not want to stay the night upstairs and we finally got two rooms on the ground floor. Fortunately it was air conditioned.

The next day we called Margaret and got directions to her place, which was in the country near Jacksonville, Arkansas. Margaret was our guide and pointed out many sites, pre-Civil War. We saw some very nice parks and a large, well-maintained zoo with many alligators, some five feet long. We didn't let the children get too close because Margaret said they could climb out of the water and come after you.

The third day I found the Cadillac dealer and checked into an air conditioner for our 1955 Cadillac. He had one of the most organized automobile shops I had ever seen. He said it would be ready next day at noon and cost under $400. Not having that much spare money and no credit cards, I called our kindly banker at National Bank of Commerce, Mr. Roy Landberg. What a time I had, though we had plenty of money in the bank.

I finally said, "Whatever, send a check made out to me because we don't like this hot climate well enough to live here."

I got the money and paid the car shop. What a joy it was to cruise around in a comfortable 65 degrees while outside it was 95 degrees with humidity up to 60 percent.

We stayed in an air-conditioned motel in Jacksonville next to the Missouri-Pacific main line. It came up from Memphis, Tennessee and woke us up every time it came by. We could hear it coming straight for the motel, it seemed, and almost the moment it was going to run over the motel the curve in the track carried the thundering diesel engines away to fade into the night.

I got up at 6 a.m, and it looked so calm and beautiful outside, I thought, "What a morning." Soon as I got out the door it was like a blast furnace, already 80 degrees at 6 a.m.

About one week in Little Rock during August was enough for all of us. We packed up, said good-byes, and headed west, traversing the same highway we came east on, but now it was comfortable driving with the windows up. We did necessary laundry, diapers or underpants, and dried them by letting them hang out the window. That was Cathy and Patricia's job. We had to turn around a few times and retrieve a lost garment when the string broke. We went back over Rabbit Ears Pass and the humidity and temperature dropped. We turned and went through Reno, Nevada, and over the Donner Pass into Sacramento.

In San Francisco, we met Mr. Deihl with Tidewater Oil Co. who was the fuel supplier for the logging operations. My, he was a fine gentleman. He spent several days as our guide and we toured several of the parks in that beautiful city and the famous waterfront park and Museum. There was history about Grays Harbor and the sailing vessels which arrived laden with good fine-grained Douglas fir lumber in a seemingly never-ending stream after the terrible 1906 earthquake. There are many two- and three-floor houses standing today and in good repair built with Douglas fir from Grays Harbor.

In 1958, Grandmother Mayr expressed her wish to go to live with Margaret in Little Rock. Marzell and I agreed there was nothing we could do to keep her here. I never did understand why she moved to town from her home on the Wishkah after Dad died.

December 23, 1958, Grandmother Brolin passed away. She had been living at St. Luke's in Centralia for six years. She was well cared for by

Brita Strindberg Brolin, 1889 - 1958.

the Sisters in her last days. She was buried at Fern Hill in Aberdeen. Father Michael O'Donnell conducted the service.

The market for cedar and hemlock saw logs deteriorated during the fall of 1958 until our company had an inventory of four million board feet of good-quality shingle cedar in rafts. No. 1 cedar logs were going for $75/M; No. 2, $45/M; and No. 3, $32.50/M, less 1 percent cash, 10 days.

I checked with cedar log mills from Forks to Raymond trying to move some cedar log rafts. We checked cedar mills on Puget Sound and finally were successful in selling large lumber logs to Seattle Cedar Lumber Co. in Seattle located near the Ballard Locks. We sold No. 2 and No. 3 dimension logs to Washington Timber Products located in Everett. The logs were loaded on rail cars at the Port Dock and off-loaded in Olympia and rafted again and towed to the mill.

During the winter of 1958-'59 we developed an order for boom logs to be shipped to Ashland, Wisconsin, for the Wisconsin Water Power and Paper Co. Their log buyer came out several times to help us attain correct specifications. The logs were cut 22 feet long and needed a minimum 30-inch and maximum of 40-inch top diameter. We loaded them in high-side gondola cars at the Port Dock, double-ending them using 50-foot inside-dimension gondola cars.

Their log buyer explained they were for towing pulpwood from Canada. On the north shore of Lake Superior, they inventoried pulp logs in bundles piled on shore. In spring break-up they would warp a set of the boom logs connected with chains end to end. The boom logs were bored on each end with a four-and-a-half-inch hole and the boom chains were about 16 feet long, with a ring on one end and a toggle on the other. They would make up a set of boomsticks like a string of wieners, then put up an oval boom.

From the north shore of Lake Superior, the boom was towed across Lake Superior into Lake Michigan and down to Ashland, Wisconsin, where the logs were loaded aboard gondola rail cars and shipped to Wisconsin Rapids, location of their pulp mill. Neither Marzell nor I were able to go to Wisconsin to see this operation. It would have been very interesting.

It was good business for us, especially during a down market. The buyer for these boom logs came several winters in a row and ordered more boom logs. We knew where to obtain spruce logs suitable for this use. We obtained most from our own operation and bought some from other loggers.

Many trees suitable for the use were found above Eudies on Highway 101. This was National Forest timber and the quality was mostly No. 2 and No. 3, which was plenty good for boom logs.

In some respects their use of boomsticks was similar to ours. Our towing, on narrow rivers with a definite channel, required a boom of logs not over 48 feet wide so we had a 44-foot tail stick, 50-foot peak sticks and 72-foot side sticks. The boom-sticks were 18-inch minimum top diameter to a 30-inch butt diameter of either Douglas fir or spruce. A swifter went across at every joint to hold the raft together.

Most of the rafted logs were hemlock or White fir and, as I said earlier, some of them would sink down in the water on the butt. To hold them we would lift the sunken end of the deadhead, tie it to a good high floater, second cut for this job. We used a flexible, single malleable wire 1/16 inch in diameter with a twist knot to hold the logs together and when the mill boomman got ready to push the log to the log slip, he would cut the wire with an axe and up the slip the deadhead would go.

From 1945 through 1968 we put up many log rafts on that little boom. Often the rafts hung pretty far out in the channel. If Rayonier had rafts tied along the bank on the other side it was a pretty tight squeeze for a tug pulling a raft down from the Rayonier booming grounds upriver.

We got no complaint from Rayonier or Allman Hubble Tugboat Co, but once we got a complaint from R. J. Ultican Tugboat Co. about the narrow gut in the channel left by our booming. Of course, if Ultican could get part of the towing, everything would work out O.K.

We didn't change anything. We knew Rayonier would never complain because they had all the river tied up for their operation. There was no channel open above the Rayonier booming grounds. This plan was installed by Polson Logging Co. long before Rayonier acquired the operation.

Bert Cole, who became Commissioner of State of Washington land in 1958, had, before that, been a logger and purchaser of State timber sales. John Pearsall, a good friend of Bert Cole, and a fellow Democrat, mentioned to me that friend Bert had a timber sale on Nolan Creek located on the alternate forties on Section 16-26-12 and, as Commissioner, was in the untenable position of issuing himself an extension on the sale. He wondered if we would have any interest, and we did. We carefully looked over the sale. It was comprised of 320 acres, with lots of good road building material on land nearby.

The contract prices were pretty high, with hemlock at $28/M, and this was before log exports. The timber was 80 percent hemlock and 20 percent cedar. Cedar was $35/M. The hemlock log market at that time was $44/M F.O.B. raft. We could come out on the cedar, but not on the hemlock. Because the sale was so far out we needed that price F.O.B. landing.

I went in to talk with Len Forrest at the Rayonier office in Hoquiam. He agreed to take eight million board feet of hemlock at the Forks reload at $43/M railhead. At that price we could come out with a small profit.

We closed the deal with Bert Cole and took over the timber sale contract. The following spring we subdivided the section so we could locate the 40s on the contract. As soon as we were ready to deliver logs, I contacted the Rayonier log buyer, Earl Simonton.

I was shocked. He said they had plenty of hemlock logs and the price was $33/M. That was the one and only time I had ever known Rayonier to go back on their word. This caused us a loss of $8/M on the saw-log hemlock.

We were able to sell the peelable hemlock to the West Coast Plywood peeler plant at Beaver for around $56/M for peelers, $50/M for No. 1, and $45/M, No. 2 so at least we came out on the cedar, spruce and 30 percent of the hemlock.

Land and timber allotments on the Queets unit were coming on the market at a fast rate and the following were main buyers: Blagen Mill Co., Anderson Middleton, Morrison Logging Co., Esses Logging Co. and Mayr Brothers. In the late 1950s, timber on the Queets unit of good quality was going at $16/M for old-growth cedar, $10 to $12/M for hemlock, $16/M for spruce, $12/M for White pine and $5/M for Lodgepole pine. Thirty years later, one-half interest in 12 million feet sold for

$1,650,000 cash, $275/M. Was quality standing timber a good investment?

Hemlock logs at the bottom of the 1950s were again difficult to sell. Green lumber F.O.B. east coast markets would net the mill around $48/M, 60 percent Standard and better, 20 percent No. 3 and better, and 20 percent Utility.

Throughout the 1950s log prices did not rise more than four to six dollars per M for No. 2 saw logs. Logs of peelable quality did fluctuate upward because demand exceeded the supply.

December 20, 1955 we received Dan, a beautiful baby boy from Catholic Children Service. He was dark of complexion with blue-grey eyes and only about a week old. By then Jennie and I were professionals with babies. On the way home from Seattle we stopped at a Swedish smorgasbord restaurant in South Tacoma. The first three children were with us. No problem at all. Now we have two girls and two boys, a good way to start the year.

Marzell and Snevah, with Tom and Mary attending St. Mary's School, bought a house on 9th Street, across from Sam Benn Park. Jennie and I, with four children, moved from Cosmopolis to 218 West 8th Street in Aberdeen to a brand new house finished in early 1959.

Dial phones went into operation at midnight, April 18, 1959, one month after we moved into our new home in Aberdeen. The new neighbors all around us were throwing cocktail parties, which is something we were not used to at all.

By the middle of the 1950s Marzell and I had been in business together for 20 years. To young people, that is a long time.

All through those years, if there was timber land for sale, we examined the property. One tract we looked at was located on the head waters of Miller Creek, which drains into the Clearwater, Kalaloch Creek, draining into the Pacific, and Nolan Creek, which flowed into the Hoh.

These were quarter-sections taken up as homesteads or timber claims and, after required improvements, filed on with lists of improvements in application for a fee patent. The timber claims, seven or eight of them, were connected by a pack train trail, the Brown Castile trail, which went from the mouth of Kalaloch Creek to the Hoh River valley via Nolan and Braden Creek.

Earl Simonton, who worked for Anderson Middleton, was with me at the time. Our companies and Grays Harbor Veneer were looking to make a timber acquisition. In several places, houses were still standing. In one I found readable newspapers, including the Toledo, Ohio, Blade (1907), the Seattle P-I (1912) and the Saturday Evening Post (1912).

I remember one house that still looked complete. It was what was called a story-and-a-half, which meant part of the upstairs had a sloping roof ceiling. Fifteen to 20 years before there had been a picket fence around front of the house where it sat back from the pack trail. The once-cleared space around the house was already growing over with spruce and hemlock seedlings about 20 feet tall.

At one other timber claim on Sand Creek, we came across another story-and-a-half house made of split cedar and under the stairway was still a bath tub made from a hollowed-out cedar log. I guess a person wanting a bath would fill the tub with hot water from the stove. There was a drain pipe in the bottom to let the water out after bathing. Several years later I went back out to the same place with Ed Maxey. By then, the house had caved in and it looked like someone had packed the bath tub out. It was a piece for a museum.

It ought to be said we did not get the timber. Charles Middleton and I drove up to Everett to meet Ned Barker, who represented the owners, prepared to offer $1.5 million dollars for the property. Mr. Barker said the property had been sold.

The total volume of old-growth timber on these properties was approximately 30 million board feet. Our appraisal, had the block not already been sold, equaled about $50/M, land included. This tract was about fifteen miles from the ocean, which made for good-quality, straight-grained cedar and hemlock. The same volume of timber located close to the ocean, like Kalaloch Creek, would have only about half of the value, especially in Red cedar. Cedar growing close to the ocean usually had extremely erratic grain structure and heavy bark seams.

Since that time the whole country has been crisscrossed with roads. Thirty years later, the area from the Clearwater River north to the Hoh is entirely roaded with wide, well-maintained roads. As a matter of fact, there is a two-lane black-top road from Highway 101 up the Clearwater Valley and across the Snahapish, up the Snahapish to the divide of the Hoh and Clearwater and down the Hoh to intersect the highway again about a half mile below Allen's Mill. All the private timber and most of the State timber has been logged. It would be virtually impossible to locate any of the homesteads and timber claims we looked over except by compass and distance measurement.

Little did I think at the time we tramped those hills that it would be logged so fast! However, the timber was over-mature, so logging it, creating all those jobs and returning great amounts of money to the State for distribution to the various trusts was a God-send to growing years of the State.

Any person traveling the black-top road and other roads in this State Forest will be amazed at the vast stands of fine young timber of all age classes from five-year plantations to 40-year-old stands of pole-size timber. Our company rebuilt and black-topped five miles of the paved road, and installed many huge multiplate culverts which had cement floors in them so the fish could more easily migrate upstream. We built the concrete bridge over the upper Snahapish River just below the honor camp, a trustee camp for the Washington state penal system.

Under the Huelsdonk Ridge timber sale contract we built three miles of mainline road on the Huelsdonk Ridge. Elevation on the ridge is between 2200 and 2800 feet. The timber was a dense stand of old-growth hemlock and Pacific Silver fir. The sale was purchased by Nichimen Company through their subsidiary, Puget Sound Log Traders.

During the 1950s, Mayr Brothers was logging and selling logs, meeting payroll of an expanding company, purchasing timber sales on both the U.S. Forest Service and State of Washington land. During this time, we had no forestry help excepting when we hired Floyd Dickenson from time to time. He was a professional forester of excellent capabilities. We knew a lot about Western Red cedar.

We always got good prices for material we produced for sales, even a simple commodity like logs need to be well manufactured. We were still purchasing timber land for our tree farm and keeping our presence on the Queets unit of the Quinault Indian reservation active.

During the 1950s the National Bank of Commerce continued as a private bank owned or at least controlled by the Price family. Maxwell Carlson succeeded Andrew Price, Jr. as President of the NBC main branch at Spring St. and Third in Seattle. Mr. Landberg was manager in Aberdeen at that time. I was always running into the bank for more money to pay off this or that. I don't think we ever got turned down.

In retrospect, looking at various times in our logging career, we ask ourselves the question "What else could we have done?" Starting with practically nothing, we were always in debt. There never really was a time at which we could have gotten out.

The year 1958 was poor for lumber producers. Blagen Mill Co., with a very large mill with a ten-foot head rig and a 60-inch Nicholson mechanical log debarker, suffered an operating loss of $170,000 during 1958. They made money with their drum barker and 110-inch chipper selling chips to Rayonier. Ed McIntosh, manager for Blagen, was good friends with Ralph Kutchera, Chief Engineer for Rayonier. This gave Blagen a good insight on one of their main customers, and Blagen was one of the main suppliers of pulp chips to Rayonier at the time they were rebuilding the Rayonier woodroom.

Blagen also was running the Picco mill located in Montesano. They were cutting dimension-type cedar logs, making tight-knot tongue-and-groove siding, sold green and used for final wall on the outside. Even though Blagen was not able to pay dividends, their reading of the market conditions was very conservative.

January 20, 1958, Marzell and I met at camp after coming in from the forest, and that evening agreed that next time the market came back up we would get out. We felt eventually a good market followed by a dull market would get us. After all, the bank only needed to call their notes and we would be finished. The best way to handle such erratic market conditions was to get out on top, and in a hurry.

The old Ned Bishop spruce mill at Junction City was purchased by the crew, and they hired Fred Maw as manager. We helped Mr. Maw get started and I am sure the crew appreciated that. Among other things, we scouted out his mill facility and figured out how we could build a log dump for him so he could buy logs direct truck delivery. We even financed it for them and they paid us back on a monthly payment schedule plus 5 percent interest. Also, we looked for people that would sell them logs. This venture began before there was any hint of log exports to Japan. Mill price for hemlock logs was still around the 1946 market, $43/M.

If round, unprocessed log exports to Japan had been stopped by the Federal government before it got started, it would have been a bonanza for Grays and Willapa Harbors. In the end the only ones that benefited greatly from round log exports were people with huge amounts of private timber.

Sawing lumber in 1958 on Grays Harbor were Blagen, Anderson Middleton, Weyerhaeuser at the Schafer Mill, Farwest at the old Bishop Mill, Wagar Lumber, and Picco's new mill in Montesano. Bay City was gone.

About September, 1958, U. S. Forest Service advertised the Sams River timber sale, 20 million board feet with four miles of road, including 100 cross culverts. The Sams River is one of the largest tributaries feeding the Queets basin. The timber was very good quality hemlock and White fir, a small volume of spruce and cedar and a small clump of Douglas fir. Five different bidders wanted the sale. We bid against Womer Brothers and got the sale for $17.75/M for White fir; $15/M, hemlock; $25/M, Douglas fir; $18/M spruce; and $12/M, cedar.

This was a very difficult sale. Soil formation on the entire length of proposed road was wet clay and situated on the north side of the hill, which means mud. Actually, the soil in the sale area would have been good agricultural land if only there were warm days to raise and dry hay. It is and always will be the best timber growing country in the world.

This was a large timber sale our company took and swallowed the over-bid by ourselves. The log and lumber market did not allow any flexibility for the mills, even though without this sale there would have been a shortage of logs.

About a mile up the Sams we found where an old-timer had staked out a homestead, or timber claim, and it was described by homestead entry survey, being in an unsurveyed part of the public domain at that time. This meant it was filed for and proved up on before there were any restrictions on the National Forest.

Marzell and his family lived on 9th in Aberdeen by the Sam Benn Park. Patricia and Mary, being close to the same age, played together and other times we had each other for Sunday dinner. The children of both families attended St. Mary's grade school. In those days, the Sisters in their black habits were doing the teaching. In the summer sometimes we would go down to Tokeland and the kids would swim in the brackish tidewater at a place called Alexander's-by-the-sea. Everybody usually had a good time.

At another time just our family went to Salt Lake City. I always liked the desert because the ground was dry and lots of rock for road building. I guess it was just an escape.

We tried to swim in the Great Salt Lake but Patricia got salt water in her mouth and eyes. It really hurt, and we had to go back to where there was some fresh water. It was so far, we never got back to see if a person could float in the salt water.

We took another trip on a mine access road into the Wasatch Mountains supposedly to find an abandoned mining camp. We stopped at a wide place and a lady came running by and said there were rattlers around the old mining camp and that a boy had been struck by a one. That was enough for us. We went back.

The next day we toured the Mormon Tabernacle where you put squealing kids in a special room and one can hear a pin drop on the stage in the main hall clear from the back spectator guest seats.

We went through the Mormon Museum which graphically portrayed the difficult life the early pioneers suffered through in order to find a new and safe place to settle. When the original settlers broke out of the rocky Wasatch range above the flat plain with fresh water roaring down from the mountains, they said, "This is the place, our God has led us here."

They built irrigation canals and ditches, and the good water, along with good soil and hot weather, grew almost any crop. That is how the great Salt Lake valley was settled in the 1850s and 60s.

The 1950s were more or less dull years in the timber business on Grays Harbor. There were no catastrophes, but a small blurp in 1951 lifted the industry out of the 1949 doldrums. Other than that short period of 16 months, the market for forest products fluctuated only with building seasons.

The exception was Red cedar. So much cedar hit the market during the 1950s, the market never recovered. We had a customer on Puget sound who bought cedar at year end to get their inventory up for tax purposes, and during those periods we shipped logs to Puget Sound. That coupled with the curtailment of logging in the Cascades in the winter allowed us to get a better price for our cedar.

During the middle '50s, we developed a cedar log peeler market with U. S. Plywood at Eugene, Oregon. William Phillips was their buyer and we made many trips back and forth between Grays Harbor and Eugene, developing a quality Red cedar peeler log grade.

Like a ship sailing away from shore into the far distance, the 1950s were gone and we looked ahead to new ideas and plans, always giving the project our best shot.

Chapter 6: The Sixties

Opportunity From the East

OUR BELOVED PARISH PRIEST, Father M. O'Donnell, as true an old Irishman as ever lived, was diagnosed with cancer of the throat during Lent of 1960. He had surgery performed but was no longer able to conduct Mass or carry on many of the responsibilities of our large parish. It fell upon Father Coyle, his assistant, to carry on and he had a severe drinking problem. This caused him a great deal of suffering and he finally ended up a complete alcoholic.

Father O'Donnell passed away May 31, 1960 at the age of 77, one of the most highly respected clergymen in all of Grays Harbor. He was our pastor during the roaring '20s, when the country was under the Volstead Act of prohibition and Aberdeen had more whore houses than churches and you could see a thousand loggers on Heron Street over the 4th of July, trading yarns about the different logging camps they were working in. Father O'Donnell was pastor of St. Mary's from those days through the Depression, the wars, and into the 1960s. He officiated through the growing-up years of Aberdeen.

It is too bad our dear mother could not have lived out her years in her own home, but at the age of 83, she was no longer able to care for herself. She took up residence at St. Luke's in Centralia, where Grandmother Brolin had lived. It was a warm and friendly place run by Sisters of the Holy Cross.

Jennie, the children and I went up almost every other Sunday and we would go to the Borst City Park. The children would play on the swings and Jennie and I would visit with Grandmother. Will that be my lot in the end or will it be like Dad, just over and gone? God knows, we don't, and just as well.

Early in 1960, we began separating our Red cedar logs from other log production. Scaling and accountability were handled on dry land using an old 20,000 pound fork lift with John Renkins as operator. We did this in limited space just north of the truck shop. Large lumber logs still went in the water if we had a firm purchase agreement from the E. C. Miller Cedar lumber people.

Land-based cedar-shake mills began to proliferate during the 1960s. Some of the earliest shake mills were Ferguson Shake and Shingle, Oakville Shake and Shingle, Carr Shake, which moved up to Neilton from South Bend. Esses Brothers was running Quinault Shake and Shingle located on the north shore of Quinault Lake.

We were producing so much Red cedar we were furnishing cedar logs to up to a dozen small mills, but we knew eventually the supply would dry up and shake mills would disappear. Other than that fleeting thought, we never analyzed the situation. We were buying cedar timber with borrowed money, so we could not salt away any good cedar timber claims, but, by the middle 1960s we still had quite a few thousand acres on the Quinault Reservation under our title.

In August, 1960, we sold Don Mackie of Mackie Shingle Co. a million board feet of shingle cedar at $70/M for No. 1, $42.50/M for No. 2 and $35/M for No. 3. His mill was on Johns River just above the highway bridge on the road to Westport. Don Mackie kept the mill after liquidation of the Mackie-Beaulieu logging operation, which was located on Johns River. They had a logging railroad, camp and cook house and logged mostly Weyerhaeuser timber located in the Johns River watershed.

Around September, cedar logs were still in the doldrums. There was so much cedar production coming in, the manufacturing market could not absorb it all, but it could change very quickly, in a matter of six weeks.

On December 20, 1960, Horace Alwood and I were down at the Port Dock preparing boom logs for shipment by rail to Wisconsin Water Power and Paper Co. While we were there, a Japanese freighter of 12,000 tons was loading two million board feet, one and two logs at a time, using ship's gear. These were some of the first hemlock logs shipped out of the Harbor to Japan. The market for good-quality hemlock saw logs was $45/M F.O.B. Port Dock, and we hoped that some day we could get some of that business. Little did we know, or anyone else, for that matter, what an impact the oriental market would have on the log and lumber business in the coming decades and how far the prices would go in 28 years.

Our cook house, bunk houses and camp had been in one location for two years, and our communications to the main office and shop at North Hoquiam still left a lot to be desired. In an attempt to get better communications, we contracted with two Coast Guard men to put in a metallic circuit line from camp out to the Highway, where we connected with Peninsula Telephone Co. which had a line down from Forks and a buried cable on Highway 101 which went by the West Boundary Road.

We furnished peeled cedar poles and had them strategically located so the lines would always clear. We finally got the line in and Peninsula Telephone came down and hooked up one of their phones for us. It did not work very well and to this day I don't think any phone will work well there until a line comes up to the shake mill at the road junction.

In a few years, we and two other loggers were able to get our own radio channels. Our call numbers were "green" and number, such as "green number one," or "green number two." The two other channels were red and blue. That gave us a fairly dependable communications system between camp and the shop, and mobile units in the various pickups and buses.

In December of 1960, we adopted David J. He was three weeks old. He has brown eyes and as a baby he had light brown hair. He was a handsome little baby and of course we all loved him. Jennie always wanted a brown-eyed boy and that certainly was our beloved David. Cathy was in junior high, Patricia in sixth grade at St. Mary's, Mike in kindergarten and Daniel was three.

We were logging two sides on National Forest and logging one side on the Reservation. We were also moving back and forth between State timber sales and Federal timber sales and down off the mountain in winter to the flat ground.

During the late 1950s and well into the 1960s, we were very active in securing fee patent Indian timber claim allotments in the northern part of the Quinault Indian Reservation. It was indeed an experience negotiating with the many owners for purchase of their timber claims. We always paid a fair price consistent with the appraisals on State and Federal lands. Our company made a good name for being fair people to negotiate with and sell timber claims to, and we never blocked anyone from use of our road system for a reasonable fee.

We planned an access road off U.S. 101 from the north about a mile west of the Kelly Ranch Road. It went through good quality hemlock, White fir and cedar along the highway, then about three miles across swampy ground covered with scrubby timber growth before dropping into Branch Creek, a tributary of Wolff Creek. In that area we found some excellent quality stands of old-growth timber. I did almost all of the scouting and, with Clarence Baker's help, ran survey lines in order to locate the various ownerships.

Some of the finest cedar that ever grew was found in the Wolff Creek and north fork of Raft River area. There was plenty of good road building gravel available and roads could be built in the winter months. We built cement bridges across Wolff Creek and the north fork of Raft River.

I don't think we ever saw more than one or two salmon in Wolff Creek and the north fork of Raft River. There never was much game or wildlife on the part of the Reservation we worked in. It was a veritable temperate zone jungle, and I think the undergrowth was so dense, the temperature so cold and the weather so wet, that what vegetation there was had no food value. One cold wet December day, Floyd Dickinson and I were camped on the north fork of Raft River and we did hear a small herd of elk. Nobody had really ever been on the land except cruisers and surveyors.

By the early 1960s, with the help of Floyd Dickinson, I had scouted out the best timber areas clear to the south fork of Raft River. One time we went in through Aloha Lumber Co. road system on the Taholah unit exploring possibilities of coming down with logs over their roads. Actually, it was faster and less expensive to haul down 101 to our yard or log dump than down Aloha road and then on the narrow secondary highway from Aloha Mill.

Timber properties had not yet attracted sufficient interest to investors and bankers to take advantage of the great opportunities that were available on the Queets Unit of the Quinault Indian Reservation. Log exports were increasing and this gave dry-land log handling a boost. The only logs that needed to go in the water were pulp logs to Rayonier. Weyerhaeuser was receiving by log yard. However, they were not consistent buyers of pulp logs.

We closed our log dump in 1966. We could sell to Anderson Middleton over their dump. Western, Farwest, Blagen, and all cedar customers except Puget Sound mills were dry-land shake and shingle mills. We also went back to Polson railroad, now owned by Rayonier, for hauling logs to their pulp mill.

We sorted our logs before loading aboard truck and marked the load as to sort. These were scaled by Grays Harbor Log Scaling and Grading Bureau before leaving the highway to the railroad. At the reload, Rayonier people put on steel strapping and swung the load aboard a rail car. When logs arrived at the booming grounds they were rafted separately for our account, sorted as peelable, saw logs, or pulp logs.

The market demand for all logs, including pulp logs, remained strong in the first half of the 1960s. We had people from Puget Sound begging for cedar logs. I think it was not so much an increased market as it was a dwindling supply of logs suitable for clear, high quality lumber, and there was probably more wood available in the Grays Harbor region than anywhere else in western Washington. We were shipping logs north to Port Angeles and into the Sound to Olympia for rafting.

The Japanese market for hemlock was steadily increasing, even though prices were extremely high compared to stateside. Log price premiums were up to $2 to $5/M for logs to Japanese buyers.

We began logging part of the old Polson Logging homestead just inside the National Forest on the West Fork of the Humptulips. The total volume of spruce was one million board feet and these were some of the largest and roughest spruce on Section 4-21-9, nine and ten feet on the butt-end. We were able to sell these huge, limby river-bottom spruce to B. C. Forest Products of Vancouver, British Columbia. They bought the whole lot for export to B.C. and we got a good price. We loaded the B.C.F.P. barge "Forest Prince" at the old American Mill dock.

The owners of the barge crane did not want inexperienced people running their cranes, so to get loaded, we needed a deal with the longshoremen.

With the help of my good friend Herb Irving, our area business agent for the IWA Union that represented our union men, we set up an informational meeting with the longshoremen, where we got an agreement with them. They issued a temporary work permit to two of our own crane operators that we paid for. During operation of the cranes on the barge, our men were present in the operator's cab with B.C.F.P. regular operators. Of course, there was no way to tell who was operating the crane. There is always another way to skin a cat.

The logs were in flat rafts and the "Forest Prince" had two large grapple cranes that could swing a log with 10,000 board feet aboard. The entire barge was fully loaded in 12 hours. The barge carried steel-hulled boom boats on the bow and, when ready to load, just lowered them in the water. These "dozer boats" pushed the logs within reach of the grapples.

After loading was complete, I went down to the dock before they set sail. The barge's tug was a wooden-hulled steam-powered ocean-going tug of post-WWI vintage. That probably was its last voyage. The logs were sold F.O.B. in rafts, Grays Harbor, accounting, scale, grade and piece count by the seller.

September 5, 1960, we had Marzell and Snevah over for blueberry pie. Norbert Charles brought us a gallon from the Mt. Adams berry fields. In that same month, I noted we should try to get out, and sell off assets. 9,000 acres of tree farm at $100 per acre would yield $900,000. 80 million board feet of Indian timber at $20/M, 100 million feet of second-growth at Kalaloch at $20/M and 10 million feet in timber contracts at $15/M would yield $3.75 million. Less debt to the National Bank of Commerce and $500,000 to G. H. Veneer, this would yield about $4 million net.

Market conditions were getting worse week by week. I suppose I should have worked 12 hours a day, seven days a week, and personally taken care of more things and covered more detail. However, I owed a responsibility to Jennie and the children while they were growing up.

About that same time, on a piece of ground along the west fork of the Humptulips, a cedar cabin about 40 feet from the river on a meadow was home for a person named Goforth, a loner. A game warden thought Goforth was poaching elk and selling the meat and sorely wanted to catch him.

The game warden, named Handron, was not famous for walking very far. On one dark night, Handron tried to sneak up to Goforth's shack without being seen. He should have known in the quietness of the forest, Goforth could hear a car coming down the gravel trail a mile away.

Down the trail through the salmon berry bushes snuck Handron. When he broke into the meadow about 100 feet from Goforth's cabin, he was shot dead with a high-powered rifle.

At a later court trial Goforth was let go free. I suppose it was on basis of invasion of privacy. Moral: Don't sneak up on anybody in the dark of night without identifying yourself, especially in the back country.

The market for logs never recovered for any length of time. The mills were constantly howling for cheaper logs and higher quality. E. C. Miller, our main cedar customer, being managed by Bob Ingram, Sr., had a nice tree farm in the upper Wishkah which they wanted to sell or trade for logs. According to Bob Ingram, they wanted to upgrade the mill, which was built around 1910 and had not been kept up. They never did get their mill modernized and ended up buying the Picco Mill from Blagens.

That was not a cedar mill at all. They had to tow their logs all the way to Montesano and haul the lumber back to Aberdeen to their dry kilns, planer and shipping sheds. In the end they just could not compete with the shake mills for cedar logs.

The Acme Shingle Co. on Elliott Slough at Bailas Dip was under Sid Lokich and Ray Sundquist. They were buying cedar logs from us. They had to open the rafts below Junction City Bridge because there was only an 18-foot opening under the bridge for passage. After pushing them through, they would brail them up and tow them about a half mile up to their mill.

They had a poor, hard-time operation. Before long, Sid Lokich skipped out to Los Angeles and left Ray Sundquist "holding the bag." They owed us $7,000 for cedar logs. Ray said he would pay that money himself and did not want to be dragged through court. He signed a personal note and finally paid the whole amount off.

In June, 1962, Catholic Children Services at Tacoma called Jennie and said they had a beautiful baby girl for us, born May 25, 1962, and we decided go ahead. In a few days we headed up to Tacoma and Catholic Children Services on Eye Street and in a couple of hours we came home with number six, Caroline Marie.

On Saturday, the boys and I went to the barber shop. We filled every chair, ages three to 13 years. Nobody said anything. They enjoyed the business. I was never ashamed of our large family and really never gave a damn what anybody thought or said. Finally I had to wear glasses for business. I could not read any longer without them.

It is a good thing I inherited the dynamic strength and patience from my mother. Adopting seven infants and trying to raise them as God-fearing men and women and being in the logging business...why, you've got to be nuts.

On January 1, 1964, Father Kevin Coyle baptized our latest addition, Suzanne Theresa, born on November 18, 1962. When Suzanne was 13 or 14, she asked me why her mother and I adopted her at our great age. It really didn't bother me. She was an easy youngster to care for and raise. I thought then, "When Suzanne is out of high school in 1981 I will be..." I didn't even want to think about that.

Cathy was enrolled at Fort Wright College of Holy Names in Spokane, in her first year in college. We took her to King Street railroad station Sunday, January 5, to get aboard the train for Spokane.

Patricia was first year in junior high. Mike, Dan and David were in St. Mary's. More and more, the Sisters were being replaced by lay teachers. Even though quality did not suffer much, discipline did, and this became a financial burden for Catholic education.

A year before, January 2, 1963, we received word of Mother passing away at St. Luke's in Centralia, God bless her soul and may He take her to His bosom. I went out to the Reservation and located my brother to tell him mother passed away at age 86. That evening Hedwig, Marzell, Jennie, Frank and I, met and discussed funeral arrangements.

Father Kevin Coyle was to officiate at the Requiem Mass at St. Mary's Church in Aberdeen, Saturday, January 5, 1963 at 9:30 a.m. Margaret came up from Little Rock, and I met her at Portland on January 3. We were all together once more.

Father Kelly actually officiated at the Requiem Mass for mother. I don't know what happened to Father Coyle.

Born January 14, 1877, Mother had a long, hard life. Her mother died when she was 10 year old, and her father drank up everything until he was broke. I was not at all sad at her death. Really, I was rather happy. She lived, firmly, resolutely and with fierce pride, a truly pioneer life. After moving from beautiful civilized Gossau, Switzerland, to a God-forsaken wilderness on the Wishkah in 1914, she often said one of the things she missed most was seeing a light from another house at night. There were no other houses nearby, and no electric lights until 1925. We had no inside toilet facilities until 1929.

At the time of Mother's death, we were moving a good volume of Red cedar peelers to U. S. Plywood of Eugene, Oregon, via rail. Bill Phillips and V. Carlson, at the direction of Mr. Bill Conklin, General Manager of U. S. Plywood, were looking for a site on which to build a green veneer plant which could peel cedar for their Eugene plywood plant. I took them to see the old American Mill site owned by West Tacoma Newsprint. Stan Targus took them down to the Humptulips River steelhead fishing. They came back with three nice fish.

Mathilda Sommer Mayr, 1877-1963. "She lived, firmly, resolutely and with fierce pride, a truly pioneer life."

Up at Nolan Creek we had a logger named Gladson logging with one side. I went to see him and pump him up a little bit so we could get more production and get the sale cleaned up. The weather was terrible, never-ending rain. The year so far had seen more than its share of winter weather, from a wild southwestern chinook down to 8 degrees Fahrenheit with ice and snow on the roads. Like years ago during horse logging, weather is a big factor in this business. A logger often stays on just to see, one more time, the morning sun come up in a flawless clear sky.

Acquisitions on the Quinault Reservation were coming to the end of the good timber. We had $540,000 borrowed from National Bank of Commerce. Forest Service timber sales would, before long, be about all of the remaining old-growth timber left for logging.

Late in 1962 we were the only bidder on Sams River No. 2, 21.7 million board feet with 6.6 miles of tough road to build with many huge, multiplate culverts at an estimated cost of $180,000.

I was active in log, timber and milling associations, namely the Industrial Forestry Association and Grays Harbor Scaling Bureau and was President of the Pacific Northwest Scaling Bureau's rules group. Most of the time I felt very uncomfortable at some of the doings, especially in the Industrial Forestry Association, which was pretty much controlled by the large corporations located in the western Washington and Oregon "Douglas fir belt."

The Department of Natural Resources offered a large timber sale on Kalaloch Creek from our road and up the east fork of Kalaloch Creek about six miles, clear to the top of the ridge where the terrain fell into Miller Creek. It was some very steep ground, all broken up, with no long ridges. Toward the Miller Creek side, the timber was larger and had more high-grade logs.

We had a total of 277 million board feet under contract in early 1963 and the export log market was beginning to move. One could already see much of the timber on these contracts would be available for export market.

People were just beginning to realize the impact of the 1962 Columbus Day storm. It would be a tremendous salvage operation. Hundreds of millions of board feet of timber were flat on the ground from the Peninsula as far south as Eugene, Oregon, on the west side of the Cascades.

We ourselves had very little blow-down, only small scattered patches, but the availability of these logs is what really started log exports to Japan. At first they wanted just ordinary No. 2 and No. 3 saw logs. They still did not have the purchasing power to buy high-grade logs. The U. S. Forest Service in the Quinault area had a 40-million-foot blow-down in the lower Sams River area. It was divided into two timber sales and quickly offered up for sale.

The logging and timber business is a hell of a career. I wouldn't recommend it to anyone. One thing is for sure. You don't want too much education, because then you would quit for sure. There are deposits and more deposits, roads to build and more roads to build, and the rainy cold season. Most all the ground in the heavy timber areas is wet clay and at this latitude and proximity to the coast you need to build all your road in about three months. The rest of the time the ground is too wet and it's like trying to plow a field in a rainstorm, the soil turns to mud and stops construction.

For three months, we are in the chips with up to $250,000 in the bank. In three months we are broke again. What a strain to run a business in such a manner. Our good friends at Rainier Bank would say it was poor management.

Everywhere, as the '60s passed by, things changed. We began to become a more liberal society. 1964 brought more changes in the logging industry and for us.

We purchased our first steel tower, a Washington Model 207 with a 120-foot telescoping steel tower on a big rubber-tired self-propelled carrier with four huge wheels. When we took the machine out new to the Matheny Creek area, I asked Father Coyle to come out and bless the new machine, which he did. It was a solemn occasion. Then the crew pulled in a few logs and we all went down to the cook house where Jim the cook had a nice little party all ready for us.

This three-drum machine could be used for skyline logging. It spooled 2,500 feet of one-and-three-eighths-inch skyline, 2,600 feet of one-and-a-quarter-inch skidline and 4,400 feet of seven-eighths-inch haulback, and it still was not enough machine for some of the long yarding distances we would later encounter. Also, we found there were not men available with experience in slack line logging. In the old steam yarder days there was much skyline logging, but those men had left the woods.

Jim Gotsis was our foreman and had some skyline logging experience from working at Schafer and Simpson camps, but we were not able to utilize the great potential of a skyline machine for many years. For some reason we did not bother to find men with skyline experience.

All log loading was now done with shovel type loading machines. Many were converted machines, but Washington Iron Works was build-

The Skagit BU 90 at Twin Peaks in 1967

ing a complete line of self-propelled log loading machines, some on rubber.

We acquired a BU90 Skagit for logging our State timber sale on Kalaloch Creek. It was a well designed machine on a carrier with tracks. This gave us two steel towers for use in logging our timber sales.

In the middle 1960s we had approximately 250 men working producing up to 3.5 million board feet of logs per month off National Forest and State timber sales and from the Reservation, with a heavy volume of cedar, still. All areas had a considerable volume of cedar, up to 15 percent.

I had wanted to hire a forester for a long time, to help me with cruising, surveying and with contractor contracts, and we had opportunity to hire Joe Ness, Mr. and Mrs. Joe Malinoski's grandson and a graduate of Washington State with a degree in forestry. He had worked for various companies and was with Grays Harbor County at the time. Marzell agreed that we could go ahead and hire him.

Timing was good for setting up a Timber and Land Department. We trained Joe Ness on how we wanted timber and timber land evaluated for possible purchase consideration, and he went right to work and set up our Forestry Department. After all, we already had 14,000 acres of timber land and were continually purchasing more as it became available. This allowed me more time to accomplish jobs I considered important. We were on many realtors' mailing lists, and we continued acquiring timber and timber land, mainly in Grays Harbor and Jefferson Counties.

The entire country was second-growth and growing into merchantable size almost faster than most people could comprehend. Twenty years later, almost the whole country was being logged a second and sometimes third time, cutting timber only 35 to 40 years of age, taking the eight-inch-and-up diameter for export. The smaller, deformed and defective logs were sold to local mills for pulp chips. We were not logging any of our second growth.

The market for logs to Japan continued to gain strength as Japan recovered from the war. A good friend of mine, H. O. "Bud" Puhn, who retired after many years as timber manager for Simpson Timber Co., took a job with one of the leading Japanese companies, Nichimen Co., and was elected President of their subsidiary, Puget Sound Log Traders. He came to us looking for logs and also said they were involved in 80 million board feet of State timber they purchased at auction in the Hoh, Queets, and Clearwater area. This was the origin of all the timber contracts which we logged and completed for them.

Mr. and Mrs. Puhn were instrumental in Jennie and I making a trip to the Orient in 1966. Mr. and Mrs. Puhn had Jennie and I to their home in Shelton, explaining all about their trip to Japan and Hong Kong. The Puhns said they enjoyed it very much.

Cathy took care of the large family of small children we had and in the spring of 1966 Jennie and I flew from Sea-Tac to Hanada, Tokyo, via 707, arriving in the early evening. We were met by Charlie Habu of

Nichimen and taken by private car to the Palace Hotel.

Jennie and I spent about two weeks in Japan and a couple of days in Hong Kong. I didn't care for Hong Kong at all. There were too many very poor people. Many wealthy people lived on Victoria Peak.

Hiroshima, then a city of 500,000, was famous and a very important log and lumber receiving port. When we first saw the city in 1966, it was completely rebuilt. Only in one place, called Children's Monument, could one see any evidence of the atomic bomb.

During the period between 1966 and 1976 we sold great quantities of logs from purchased state timber sales to Japan. In the case of Nichimen, we took title to the timber upon cutting the timber. This method of doing business gave us the kind of contract we needed, but operating on such a large scale required far more capital than we had available. Buying timber required large amounts of cash on deposit for performance bonds and credit on felled timber.

In January, 1968, Jennie and I adopted our eighth child, John Richard Mayr, born December 29, 1967. John was dark complexioned with brown eyes, a very handsome boy. He fit into the household. Patricia went to work in 1968 and moved away from home.

In May, 1968, it looked like a very severe fire season was ahead of us. June 1, on a Saturday, it was hot and dry, and, at home, I heard of what became known as Sams River fire.

Anderson Middleton had a timber sale on the east side of Sams River where they had Buswell Brothers building road. They already had a bridge in place for crossing the river. On that hot, dry day, they were burning right-of-way slash in a controlled burn and an east wind came up and pushed the fire across Sams River into our felled and bucked timber.

The fire ran through our entire cutting unit, damaging all of our logs. It charred our Model 207 Washington, burning all the sled runners, the cable and all rigging in the tree. The Washington TL6 Log Loader was also crisped. All the logs on the landing needed the charred ends bucked off.

We had never been through a fire that damaged our property before. The insurance for the contractor working for Anderson Middleton made an agreement with our insurance company. We got a fair settlement and salvaged what we could.

Besides burning the felled timber, and destroying our equipment, the fire killed about 10 million board feet of green timber. We got one of the sales and Don Bell got the other.

On July 16, 1968, a fire broke out in the Quinault Indian

Reservation. This fire was caused by friction of the haulback on a logging show being carried on by Esses Logging Co. Jennie and I, with our family, were at Disneyland. On the way home we stopped at the Columbia River at Kalama Inn. I looked at the news stand and in big headlines was the report of a Quinault Indian Reservation wild fire which burned over 12,000 acres.

Not knowing the exact location of the fire, I called Evar Carlson, our office manager. He told me everyone was out on the fire line and he still had no knowledge as to the exact limits of the fire. Later I found we were very fortunate. It never reached our operations or property, even though we were located to the northwest, the prevailing wind direction that close to ocean, only two miles away.

It was a nasty fire to bring under control because of the great amount of cedar slash and large number of cedar snags in the uncut areas adjacent to the logged areas. The fire burned into the Crane Creek Unit and part of the Taholah Unit, areas under contract to ITT-Rayonier and Evans Products, respectively. Between the Sams River fire and the Reservation fire, our crew, as well as crews of other loggers, spent about two weeks fighting wild fires that year. We had not been on a major fire since the big Forks fire of 1950 and the Bear Creek fire north of Forks in 1951.

I got in on both those fires, and the Forks fire was a monstrous thing. It covered an area from Highway 101 easterly from Forks for about twelve miles and on each side of the old Peninsula and Western Railroad grade, which by then had been abandoned. It burned up from the Calawah and Sol Duc Rivers east towards the higher ridges.

The Forest Service headquarters fire camp was at the Snider Ranger Station.

I was on the fire with about 10 of our men building fire trails and falling old snags. I slept in a carryall panel which was awful cold toward morning and not comfortable. Most of the men slept on straw laid on the ground in an old warehouse. The portable kitchens set up by the Forest Service turned out great food, including sack lunches which were sent out to the fire crews.

One part of the fire that I remember clearly was on a plateau in a 40-year-old second-growth Douglas fir stand. There were many old early-logging-day logs lying about and everything was tinder dry. The fire roared through the second growth. It would jump up a tree and crown, the flames a-roaring, the dead limbs gone. There was not much a person could do.

Later we backed away from the river, built a wide fire trail with a cat, and cleaned up all the dry fuel, which, after it burned, gave us a controllable fire line. With hundreds of men working under experienced Forest Service leadership, we gradually worked the scope of the fire down and men and bulldozers started mopping up.

Another area we were sent to was high up on the ridges. I knew it was at least 2,800 foot elevation because there was some beautiful White pine in an old-growth stand. It was so far up the hill it took three hours of walking to get to the fire line, and about the time we got there, cloud cover came in and it began to mist and by next morning the fire was really laid down. It was late September, so we were released from duty and allowed to go home and back to our jobs.

In researching the Forks fire, I contacted the Forest Service Supervisor's office in Olympia, asking for information. They told me the information was in the archives, and they would try to send me copies. I never heard a word back from them. I also contacted the state Department of Natural Resources in Forks. They said they weren't in operation at the time of the fire and had no information.

In October of the following year was the Bear Creek fire. It was in a northern part of the forest that had been railroad logged and apparently burned several times before. I had eight cutters with power saws and tools and we were instructed to fall snags.

This was a poorly organized operation. Off to the north, we could see camp fires made by deer hunters while we were trying to control a wild fire. That didn't go over worth a damn. At the end of the drivable road was a bus-load of supposed fire fighters who had on tennis shoes. It looked like they had been recruited by one of the logging companies to fill their quota of manpower. It was clear to see the only fire fighting they would be able to do was putting out cigarette butts.

I got together with our crew and said, "Let's fall all the snags from here to the far ridge and then we will take our tools and get the hell out of there."

We commandeered some of the guys from the bus to pack tools for the cutters, and in one day our crew did more than the bus crew could ordinarily do in a week. Late in the afternoon we had all the snags cut and put the tools in the bus and and headed for home.

It was cold and damp and felt like rain and before we got back to our camp on Salmon River it was pouring. I knew, though, if it hadn't rained, I would have been in a hell of a lot of trouble. You don't just walk away from a wild fire, but by the time they caught up with me it was a moot question. Even if they had given me trouble, somebody would have gotten a good going over because the whole project was an inefficient and wasteful operation.

In 1967 and early 1968, the log export market took a jump of over $100/M. Demand was strong, prices on No. 2 hemlock was getting close to $150/M F.O.B. dock and we had just made some good timber purchases on both State and private timber stands.

Jennie and I made another trip to Japan as guests of Pacific Lumber and Shipping. Their main customer Shin Asahigawa and the Nichimen people just couldn't do enough for us. That time, we went as far south in Japan as Kyusho Island. They wanted us to see the steaming geysers and hot baths. We never did find their hot baths hot enough to be comfortable, even if you let the faucet run for half an hour.

1968 was the best profit year the company ever had. When a person looks back twenty years then a person wonders, "Why didn't we get out?" The market was, and still is, like a yo-yo, up one time and, in less than six months, down again.

Congress was coming under pressure to put log export restrictions on National Forest timber, but not enough pressure was on log prices yet. Our representative in Congress, Julia Butler Hansen, put legislation together, effective two years later, restricting from export round logs originating on National Forest land.

It was approved by Congress and signed into law by the President and soon had the effect of raising stumpage values on both private and public timber sales, though it was a restriction on only a part of the natural resources. That set the stage for two timber values. Non-restricted timber became more expensive than restricted timber.

Eventually, this led to legislation and restrictions which, in effect, denied exporters of private timber access to bid and qualify for National Forest timber. For an operator to continue "wooding" his mill, he needed National Forest timber.

The cost of private timber often was more on the stump than the market would bear on finished lumber. You could sell logs off unrestricted land for more than the market price for kiln-dried, finished lumber. If you could make lumber out of unrestricted logs by merely waving a wand, your lumber was worth less after the wand waving than the logs were before. Monte Dahlstrom told me he knew of no business where you could take round logs, manufacture them into green veneer, and the veneer was worth less than the logs were.

As Harry Truman said, "If you can't stand the heat, get out of the kitchen," but it did not go that way. All the kitchens were shut down. These were black times for us.

We wanted to build a mill. No new mill had been built on the Harbor for many years except Picco Mill in Montesano, but that was for cutting

Scenes of a "bottom" loading green hemlock for the east coast became a thing of the past when the Calmar Lines stopped hauling Bethlehem steel to the West Coast. Photo by Jones Photo Co.

green hemlock, and hell, they couldn't even get a ship to Montesano unless its draft was 11 feet or less.

Our idea first was a salvage mill to cut the good stuff out of defective logs. Then Marzell and I talked about how we could not find any kind of halfway decent salvage mill anywhere.

Then came another jolt. The Calmar Line, a Bethlehem Steel subsidiary, discontinued bringing steel from the east coast to California. They had been coming up the coast to pick up green hemlock lumber at northwest ports for return trip to Brooklyn, Boston, Philadelphia, and New York.

This was the death knell for Anderson Middleton, Western Lumber, Farwest, Blagen and Picco. Neither the manufacturers nor the brokers on the east coast could get a bottom (ship) of a foreign flag, not even a Canadian carrier, because of the Jones Act, which guarantees U.S. shipping lines all cargo from U.S. port to U.S. port.

The Jones Act had a tremendous effect on lumber manufacturing in the Northwest, and it added fuel to the log export industry. A buyer for foreign market could use a carrier of any flag, whichever they could get chartered for the least money.

The Canadians jumped right on this opportunity and geared up to grab the east coast share of the green Hemlock market. Maybe one or two U. S. vessels came in for a while, and then there were no more for intercoastal trade.

You sure as hell can't ship green hemlock lumber 2,800 miles and over two mountain ranges by rail and be competitive. With green hemlock, surfaced two by four inches, you can get 12,000 board feet on a standard truck and trailer and be at 74,000 pounds. Kiln dry lumber to 12-14 percent moisture, and you can get 26,000 board feet on the same truck and be legal.

Our company continued to expand and take on more payroll and added responsibilities. Marzell wanted our own road building operation, including at least five bulldozers, several shovel loaders and all our own gravel trucks. Our payroll was up to 260 persons during summer months when miles and miles of logging roads needed to be built in order to reach timber in the higher hills of both the State and Federal forest.

In the late 1960s we were building road at the head of the north fork of Matheny Creek in what looked like huge, beautiful white wood forest running, we thought, 100,000 board feet to the acre. The defects in those huge trees cut the net down to 35,000 board feet per acre, resulting in proportionate costs on road building and logging increasing 150 percent.

On the defective timber, we contracted the cutting out to Northup and Warren, who had contracted with us for several years. On a fateful Saturday in August, George Warren was falling, and August Northup and Warren's son were bucking.

Warren felled a big hemlock, about six feet, diameter breast height, and August and young Warren proceeded to both buck on the same tree. Warren proceeded to put an undercut in his next tree, a huge White fir. He had hardly scratched the tree and it let go, right down parallel to the tree August and young Warren were working on. Both were killed instantly. What a terrible day.

Ralph Blaine, our foreman, was in camp and helped George Warren all he could. The shock of this terrible tragedy lingered with me for years and years. Injury and death in the logging game is so violent and so damnably final.

The defective timber caused us problems in marketing the logs. Some of the logs we considered peelable were shipped in over the railroad, scaled on rail cars, bundled and then rafted in bundled rafts. At one time we had four rafts, about a million board feet, ready to ship invoiced to the Grays Harbor Veneer Corp. They waited until the invoice was due before telling us the logs were too defective and they did not want them. That was, I thought, a hell of a way to treat us after 20 years of being a major log supplier.

At one time A. J. DeLatuer said they would sell us the Grays Harbor Veneer operation. They had about 600 acres of old-growth timber on

Raft River and the old veneer plant always made a little money, but all of a sudden they had no further interest in talking with us. At the time, we thought the owners, A. J. DeLatuer, L. Cole and Puget Sound Bank were planning and negotiating with Anderson Middleton. Wayne Hagen, an assistant manager at the G. H. Veneer plant, was also friendly with R. W. Middleton. It was not too much later we found out Anderson Middleton bought the G. H. Veneer operation, so everything fell into place and we were on the outside looking in.

We were coming to the end of the 1960s, and many changes came about in our industry. The saw-milling industry was almost nonexistent on the Harbor except for Weyerhaeuser who was operating the Schafer Brothers Mill located in South Aberdeen. It had a boiler and dry kilns which allowed them to market dry hemlock lumber by truck or rail shipments.

The round log export market continued to expand every time the market hit an upward trend.

The plywood mills were beginning to use peelable hemlock logs as underlayment in manufacturing of plywood panels, and we were able to sell peelable hemlock logs to West Coast Plywood, Harbor Plywood and Hoquiam Plywood.

For many years, we looked for a long-term land bank that we could use in financing our timber land acquisition program. Through my friend Siebert Larson, a potato farmer from Elma, we were introduced to Mr. Whittaker, manager of the Elma office of Federal Land Bank of Spokane. He told us they were interested in making long term loans on land with marketable timber.

Mr. Whittaker explained the land bank was an agricultural lender set up by charter from the United States under which farmers could group together and form a long term lending institution. At its inception, the Federal Government prepared a Federal loan for the new bank. That loan was soon paid off and the bank was operating on its own credibility selling securities on the New York money market and enjoying the same interest rates on like-term paper as the Federal Government.

In 1967 we negotiated our first long term loan with the Federal Land Bank of Spokane for an amount of $550,000 at 5.5 percent for 30 years. Interest rates were stable and remained so for many years. This gave us about a three percent savings on money used for our main raw materials, trees and timber.

The Land Bank could only finance property in which clear title for land and timber on the land was an asset pledged by the borrower. It was required in the case of a loan to a corporation that the borrowers pledge their own assets as guarantee for the loan. Even in those times of minor inflation, money at those rates used to purchase established timber land with merchantable trees growing on the land made for a sound financial risk, to the lender and the borrower.

We used the new source of capital to full advantage in an arrangement with our commercial bank, National Bank of Commerce of Seattle, through our local branch. This enabled us to obtain quick funds to make an attractive purchase. Our local bank would hold the paper for up to two years or until we had a sizable block acquired that was not for immediate liquidation and would fit into our timber land portfolio pledged to the Federal Land Bank of Spokane.

In those times the sun was shining and the sky hung with violins playing sweet music. Several times, Jennie and I, while in Spokane, called on the Federal Land Bank main office. We were treated very well and apparently they thought highly of us, but I could detect the steely, fishy-eyed look of a professional banker, which, liberally interpreted, meant, "So long as you are well and make your proper payments we have a partnership between us. But heaven or hell can't help you if you get in financial problems. We'll cut you up like a fisherman cutting up bait."

During the summer of 1968, my friend A. M. "Mac" Polson mentioned his family had about 4,000 acres of timber land they would like to sell. Most of it was in western Grays Harbor County, and a few pieces were up on the Clearwater. Some on the Clearwater was undivided with W. C. Plywood land purchased by U. S. Plywood and subsequently by Nissho Iwai. We were able to negotiate a firm price around December, 1968, after their attorney drew up the papers. All the papers were in order and we had the funds already in hand, having arranged a loan from National Bank of Commerce.

The Polson attorney called me just a few days before the closing date, saying he had not seen Arnold Polson for some time and could not verify his signature. In a day or two, Mac Polson called me and said Arnold Polson had passed away and that it would be late January or February, before they could furnish merchantable title.

I knew Arnold Polson from the years during the war when he was manager of Polson Logging Co. Their company hauled logs for us from the Lake Quinault area and boomed and rafted them. He was a large man, about 6-foot, 6-inches, and seemed bitter and very hard to talk with.

In one incident, years before, the Forest Service wanted access over Polson's abandoned railroad grade from Highway 101 northeast to the

National Forest boundary on the west fork of the Humptulips River. Polson Logging Co. did not want to negotiate so the Forest Service brought proceedings to condemn the property. The Federal Court awarded the abandoned right-of-way to the Forest Service for $25,000. In a negotiated deal, Polson could have received considerably more.

Moral: Don't fight the Federal Government on their access needs for an abandoned road.

The timing was right for us in the Polson purchase. We were able to start logging within six weeks out on the Walker Bottom Road, also on the Newskah, with a subcontractor and later in the Hoquiam watershed. As we completed logging various Polson tracts, we always replanted the land.

In the summer of 1969 my good friend Floyd Dickinson asked Mike, Dan and myself to go up to Blue Glacier on Mount Olympus. That sounded like a neat idea and Mike and Dan were eager to go, so we agreed to meet Floyd at the jumping-off place where the Park Service has a parking area at trail head on the upper Hoh River.

There we met Floyd and he had two fine looking horses from Mini Peterson, who lived on the Hoh. We finally got everything packed and on the horses and started up the trail about 11:00 a.m. We walked in about 10 miles, leading the horses with their packs. It was a nice easy trail all the way through old-growth stands of huge fir, spruce, cedar and hemlock. Wherever a huge tree had fallen across the trail it was bucked out wide enough to let a pack horse through.

The trail evidently had been established before the Olympic National Park was formed. In all the areas that were swampy there were corderoy puncheon laid to keep out the mud. After five hours we came to a beautiful meadow, about 25 acres in size, and Floyd had us unload the packs and make camp for the night.

We made a little camp, tethered the horses and Floyd made a first-class, camp-fire dinner; steak, baked potatoes, mushrooms, salad and all kinds of good stuff.

After such a fine dinner everyone slept good, and the next day we were up at daybreak. We made light lunches and the boys took the first turn riding. Every mile or two we would trade off.

The Hoh Valley rapidly became narrow and the hills went up a much steeper pitch on both sides of the valley. Streams came cascading down and the trail became more rocky, with more big boulders, more curves and sharp pitches in the trail. We crossed a well-built log stringer bridge about 60 feet long over a deep chasm that had a swift, roaring stream even too rough for fish.

A Mayr Bros. truck on a Mayr Bros. constructed road.

We were probably at about 5,000 feet elevation, and the timber became smaller and more limby. The trail hugged around a steep hillside of about 100 percent slope and it looked like at least 1,000 feet to the bottom with nothing in between. It was so steep you could reach out while sitting in the saddle and touch the upper slope cut.

After about an hour of this, we broke into gentler ground that was more open, and we could see ahead that it was not far to the Blue Glacier. Finally, we came to a little shelter in a clump of mountain hemlock and White fir. The trees were only about six to eight inches in diameter and 30 to 40 feet tall. We tethered the horses, ate lunch, and went on to the Blue Glacier on foot.

In a very short while, we were on hard packed snow, probably at around 6,000 feet elevation. It was a clear and cloudless sky, and it seemed like only a short distance to the top but Floyd said we were not prepared to go that far that day. After walking a distance, Floyd said we should not go too far because we had no climbing gear and there could be crevices in the ice. We would need a full base camp at the clump of trees where we left the horses if we were going to climb the mountain.

We went back down the trail to where the horses were tethered, picked them up, and high-tailed it back to our base camp. It seemed like it didn't take near as long to get back as it took to go up.

Floyd fixed a nice camp-style dinner; rice, raisins, salami and bread. Everybody slept good. Next morning, bright and early, we headed back to the jump-off point. We arrived at the truck around 2:00 p.m. and helped Floyd get the horses aboard. The boys and I arrived home around 6:30. Mike and Dan will remember that trip all their lives.

The next year, Mike, Dan and I went up to Floyd Dickinson's cabin on Beaver Lake, B. C. It was in June and it got up to 95 degrees, so hot nobody minded wading out through the algae to get into the lake. We fished Beaver Lake and for the first time went on a look-see on McCauley Lake.

There was no decent place to launch Dickinson's boat at McCauley, so we went on a farm road all the way around the lake to an old farm that had a little dock. Dan and Mike fished out of the boat, and I tried fishing off the dock but the mosquitoes were too ferocious along the shore of the lake. We were glad when Mike and Dan came back. We put the boat back in the station wagon, headed back to the cabin and went swimming. Then the air-conditioner quit in the station wagon so we decided to pack up and head for home. After we got on the road it cooled off rapidly.

The following November, Mike, Dan, Dave, and I took a drive up to Williams Lake and stayed at the Slumber Lodge Motel. The next morning, we headed out to Beaver Valley to look at some property that was for sale, about three quarters of a mile below Dickinson.

We got on the wrong road and headed for Horsefly, and finally came to a left turn. We took that and, after a few miles, stopped at a little farm house, and asked the lady if this was the Beaver Valley Road. She kind of rolled her eyes and said, "This is Beaver Valley." It was winter and would get dark early. At least we knew where we were then. At that time, there was not a identifying sign on the whole of that road.

We found Dickinson's cabin and located the 15 acres along the lake. It was nice level land with gravel under-strata and almost completely timbered with fir, spruce and aspen. It was a nice piece of property and, before too long, we purchased it for $6,500, Canadian. At the time, Mike was about 14, Dan was 10 and David about 6.

Little did it dawn on me how they would change in just a few short years. This was about the time of Vietnam and some of the goofiest philosophies to ever hit America, including crazy music, pot and all kinds of crazy ideas. Even educated people were hung up. It affected all of our boys, but none of the girls.

The '60s weren't all work and no play. These are photos of a fishing trip on the Upper Queets with Dan, Mike, and their cousin Tom in 1966.

It did in a few years turn a happy family into one not quite so happy, but somehow we worked our way through. Jennie and I made a commitment which to us meant we must continue doing our very best, but when you see 14- to 17-year-olds in drugs and alcohol you wonder what went wrong. Apparently trying to raise children under Christian influence didn't do the job at all. However, they all graduated from high school on time, and Mike attended college for one year.

Looking back on the years 1960-'69 is not much different than looking back on any journey in life through uncharted waters. Even people who do things and accomplish much are usually not remembered as being particularly intelligent, successful, or whatever the banker calls successful, unless they accumulate large amounts of cash and deposit that in the banker's bank.

We were coming to our 35th year in our business, it was neccessary for me to spend more time in the office running the business. Marzell was taking all of the field responsibilities and had a good organization of supervisors, including Nick Disho in cutting crew, Jim Gotsis, Bob Burbank and Mel Brooks in the yarding and loading, and Urho Wirkkala in road construction. In the shop for maintenance and upkeep of all equipment we had Horace Alwood and Ernie Hansen.

Down through the years, log hauling became an increasingly important aspect of the log business. Transportation was often more than 35 percent of the cost of the entire logging operation. Some log hauls to the upper reaches of the National Forest and the State Forest in the upper Clearwater were up to 150 miles, round trip. A huge investment in logging trucks was required to move 300,000 board feet of timber daily from the stump to the dump.

The 1960s were the peak of development and production of the last large old-growth hemlock forest in the lower 48 states. Logging these thousands and thousands of acres of old growth timber on the Quinault Indian Reservation, and in the State Sustained Yield Forest Number One and the Grays Harbor Sustained Yield Unit of the Olympic National Forest was just another facet in development of the west.

Although it created no continuously sustainable economy, at least cutting all this timber finally did prepare the land for a new crop. Perhaps in the year 2040, there may then be a lumber economy supported by fine stands of second-growth Douglas fir.

Chapter 7: The Seventies

The Sky's the Limit

THE 1970s WAS THE PERIOD of the greatest inflation of the postwar years, brought on by the oil cartel. Stumpage price in the decade increased four times. Hemlock rose to between $250 and $350/M and old-growth cedar increased from $100/M to $800/M. This inflationary pressure was brought on by the strong demand for logs by the Japanese market and, to a lesser extent, the new export market into Korea. All the while, demand was also strong for forest products in the United States.

In 1971, Marzell and I made the decision to build a sawmill beside the Hoquiam, up-river from the shop, between the chip mill and the bend of the river.

Ray Aarhaus, superintendent of our drum barker chip mill operation, was put in charge of planning, and we began the planning from the large old-growth hemlock and White fir premise. The mill's production would be all types of lumber, usually reserved for a cutting mill.

We began with 60-inch mechanical ring debarker, an eight-foot head rig, and an all-steel building set on piling driven to refusal. All machines sat on mass piling with concrete and reinforcing down to the moisture level. The land included in the mill and log-yard plan was flooded at high tide, so the entire site was filled to a point of two feet above the highest known high-tide mark.

Ray Aarhaus used one of the houses across the street from the main office for his planning office and, after thorough research, built a complete scale model of the mill.

After Ray completed the pilot model of the mill, we called in Jerry Crow from Crow Engineering of Portland to have his firm of draftsmen and engineers make complete construction drawings,

Tim Lockey with old-growth spruce logs in the yard at our North Hoquiam mill. Photo © John Tylczak.

including machine spacings, location of drives and chains, sizes and speeds. We began construction in April, 1972, using Franciscovich Co. as lead contractor with men on our own payroll filling whatever jobs they could handle.

Timing was good for building a sawmill on Grays Harbor. It was the first new mill built on the Harbor in years, and construction went on rapidly.

By the time the mill was completed Ray already had some of his key people working for our company, including the saw filer, head sawyer, twin band operator, and edger operator. An even one year later, on April 1, 1973, we cut the first log. It was a two-foot diameter hemlock.

Darrell Barnes was on the controls, and it went smooth as silk. Edmund Pomranky took over the controls and became the head sawyer, George Kugan head filer, Ron Webb second filer. The mill ran one month on one shift and then went on two shifts and remained on that basis most of the time.

Log supply for the mill came mainly from Hook Branch Ridge, the main Matheny Ridge and one unit remaining on Dilly Ridge. This last cutting unit was contracted out to Mike Tobin during the early 1970s. We sold the first production rough and green. However, we followed rapidly with a Kipper & Son, 40,000 pounds-per-hour, all-waste-wood-fired steam boiler, with maximum pressure of 400 PSI, and at the same time built three dry kilns with cooling sheds.

Again, our timing was good. We were the first hemlock mill on the Harbor with dry kilns since before the Depression, except the Schafer Mill, now owned and operated by Weyerhaeuser Co. Green hemlock lumber was history in the domestic market because of the Jones Act, which prohibited foreign ships from hauling freight between American ports. There were no longer U.S. flag vessels available to carry hemlock lumber from Grays Harbor to the Atlantic seaboard, which was the only domestic market for green hemlock lumber.

There are several standard measurements.

1. American Standard Surfaced, either kiln-dried or green: It takes a sawn two-by-four inch piece to make a kiln-dried one-and-a-half by three-and-a-half inch "two-by-four." It takes a two-by-six-inch piece to make a kiln-dried one-and-a-half by five-and-a-half inch "two-by-six."

2. Full Sawn: net two-by-four, two-by-six, three-by-eight, or four-by-four inches.

3. Lumber for millwork, measured in quarters of an inch: 5 quarter (one-and-a-quarter inches), 6 quarter, 7 quarter, and so on.

4. Lumber sold overseas (export): The order specifies and is measured full size, four-and-an-eighth by four-and-an-eighth inches, or three-and-nine-sixteenths by three-and-nine-sixteenths inches, and are specified to order.

Overseas to Japan for green four-and-an-eighth by four-and-an-eighth and United Kingdom for two-by-eight- and two-by-four-inch door stock green was a market. Also, United Kingdom bought six-by-twelve-inch by 14 feet and longer timbers for use in the City of London rebuilding centuries old buildings.

Our lumber sales were through Pacific Lumber & Shipping of Seattle. They assigned Mr. Chelsea Browne to represent and handle our lumber production. Our lumber was well received by lumber customers throughout the marketing area.

Our son Mike graduated from Aberdeen High in 1972 and went to work for the company in the log yard, learning about log grades, quality and export sorts. Daughter Patricia worked on computer key punching under Jim Reynolds. Our Cathy married after two years in college.

Marzell's daughter Mary went to work in the office handling payroll and accounting entries in 1973, after she graduated from Marelhurst in Portland with a degree in Home Economics and Business.

Tom, Marzell's son, graduated from Oregon State University in 1974 with a degree in Forestry. Tom went to work in the forestry part of the business which included timber sale purchases, timber land acquisitions, and charge of the forestry staff dealing with subcontractors in falling, bucking, yarding and loading

During 1973-'74, the market for lumber was very strong and brought very good prices. This market hung on until the 1974 world-wide oil crisis, caused when foreign oil-producing nations joined together and formed an oil marketing cartel. Crude oil prices increased fivefold within one year to over $30 per barrel. This caused a world-wide scramble for alternate fuels, and more efficient automobiles. Many new supply sources were found throughout the world, and within three years oil prices swung the other way.

As world-wide economic structures adjusted to the higher oil prices, lumber markets began to recover and, by 1977-'78, overseas market for logs and lumber was again strong, pushing up prices again.

Our company was actively involved in log exports. Before 1973 it was allowed to export National Forest logs, but after we were ready to start our mill, that was no longer permitted. Our export logs came from Indian timber sales and State of Washington Department of Natural Resource timber sales. Our log export volume came to 4 million board feet per month.

In 1973, we learned that Clive Abel was seriously ill, lying in a Tacoma hospital. Clive was born to W. H. Abel, a brilliant early-day attorney from Montesano. While he was in the logging business, his company was M-D logging. A good friend of mine said it stood for "mine and Dad's" but I knew the name M-D came from former owners Malinoski and Duffy, who logged spruce in the North River area during and after WWII.

Clive's wife was able to find a judge who granted her durable power of attorney, which apparently gave her power to sell and pass clear title on any of Clive's property and she wished to sell property. The total acreage was about 1,000 acres of 90 percent old-growth hemlock, White fir, a few Sitka spruce and 3 million board feet of Red cedar, for total of 40 million board feet.

Floyd Dickinson and I made an analysis of the property and found that it was far enough inland that the trees were well formed with a minimum of swell-butt logs and the defect average was low. I instructed Floyd to act as agent for us and to offer her $650,000 cash. That amount of money was a shock to Mrs. Abel and she very quickly approved the sale of the property. We paid Floyd Dickinson a negotiator fee of $10,000 for bringing us together.

At the suggestion of Mrs. Abel, we met in Clive's room at Tacoma General Hospital. I was surprised at his appearance, having not seen him for several years. His doctor told him years before to quit his excessive drinking.

Poor Clive, lying on that bed, was a sad sight. He could not move a muscle except his eyes and eyelids, the massive stroke had so devastated him. In our presence, Mrs. Abel asked him something like "You don't want to sell the land with the timber, do you, Clive?"

She then interpreted a roll of his eyes as "No."

Mrs. Abel's attorney was also present and I think now the purpose of the question was a round-about way to let Clive know she was selling his holdings. The purchase price of $650,000 was agreed upon. Clive's wife was so eager for the money, she asked her attorney if he could get the court to approve the sale that day so the $10,000 deposit we made could be released to her for "spending money" for her and companion.

We closed the deal and a year or two later we were able to acquire the land. McKay and son did most of the logging, which was completed in three years, after which we aerial seeded the entire logged area.

In January, 1975 we were contacted by the Polson family. Arnold Polson had passed away leaving his widow and two daughters. They had 40 million board feet of very good quality old-growth Red cedar with a mixture of hemlock and White fir on patented allotments on the Quinault Reservation. Much of it was undivided one-half interest ownership with the other ownership held in trust by the U.S. Government for allotees. Short & Cressman were handling the sale for the Polsons.

The cruise volume was 30 million board feet of cedar and 10 million board feet of hemlock, and the asking price was $5 million, which was a fair price at which a person could make money. The land went with the timber.

After we completed the negotiations with Polson family, I approached National Bank of Commerce through Mr. Roy Landberg, for a $5 million loan which was quickly approved.

The purchase was consummated in the Short & Cressman office. Mr. C. W. Adams was attorney for us.

Some time after this transaction was complete, I heard from my friend, Stuart Ferguson, that Jim Jackson, lately retired Chief of the Taholah people, was very irate and disappointed when we made this purchase from the Polson family. I knew full well that, even with all the other holdings we had on the Quinault Reservation we still needed Jim Jackson as a friend.

I had known him for many years, and with that in mind, I asked Stuart Ferguson to try and arrange a meeting in his office at Aloha, Washington. Stuart was able to arrange such a meeting and Jim and I did meet.

Jim said he was irate because he was negotiating with the Polson family to purchase the holdings we bought. I said, "Jim, I knew nothing about such negotiations. We were approached by Short & Cressman, attorneys for Polson family, to see if we would have any interest in the property." We understood each other to be truthful. Jim accepted my explanation. We shook hands.

Jim Jackson and I always had a good rapport. I knew Jim's father from many years before when he was professional timber cruiser, a very dependable person.

Jim's son, Cliff, had a complete logging outfit and was looking for work. We negotiated a logging agreement with Cliff on five of the 80 acre parcels from the Polson purchase.

On four of the allotments we bought undivided half interest from Polson family in fee simple, the other half interest was owned in trust by the Snell family and controlled by Elizabeth Cole, Quinault tribal member.

We negotiated the following agreement. On a basis of one allotment at a time, we would agree on the value of the timber on both our half and the Cole half. Then, Elizabeth Cole would get a one year cutting permit on all the timber on the 80 acres, and our attorney would prepare a deed for our half interest in favor of the U.S. of America, held in trust for Elizabeth Cole, et al.

We would close the deal in the B.I.A. office in Hoquiam in the presence of Superintendent of Indian Affairs for the Quinault Reservation. In the transaction Elizabeth Cole, by instrument, would hypothecate her cutting permit to us, approved by the Superintendent.

Elizabeth Cole, et al., got their land back in trust, and we received the timber rights. We would hand the Superintendent a clear title deed to our half interest and Elizabeth Cole, et al., a check for their half interest, and we were ready to proceed with harvesting the timber. This worked very well and we received full cooperation with the native people.

This later had a favorable impact on our ability to acquire the Taholah contract from Evans Products Co. This was for completion of a very large contract Aloha Lumber Co. had purchased 30 years before, over 900 million board feet of large, good quality Red cedar, located north of the Quinault River, west of the Crane Creek unit and south of the Queets unit.

Access into the unit was from Aloha Mill via private road north across the southern part of the Reservation and across the Quinault River via a unique suspension bridge, 450 feet long from bank to bank, held up by two-and-a-half inch cables.

The suspension bridge was the only dependable way to cross the Quinault River. It was designed by Mr. Millward, long time logging engineer for Aloha Lumber Co. The shore piers were cluster piling, well in on the shore, away from danger of wash-out by extremely high water. The cedar timbers supporting the frame of the running deck were huge, 18 inches by 20 inches, 20 feet long. The bridge was designed to support a 100-ton load and remained in place for 35 years, suffering only minor damage at the piers from high water and an occasional cracked or split timber support that needed replacing.

Our new mill was operating in a very efficient and satisfactory manner under supervision of Ray Aarhaus. I was no longer going to State timber sales. Marzell was doing the bidding. Tom was doing the appraisal and bidding on B.I.A. timber sales, and I was still planning bids on Forest Service timber sales. We were logging National Forest on Salmon Knobs, a very nice old-growth White fir and hemlock sale on Salmon River Ridge. We acquired another quality sale, Salmon Dehorn, located in the same area of Salmon River Ridge.

We were allowed to purchase 12.5 million board feet from the Forest Service with an export limit of 22 million board feet of private timber. With the great quantity of Indian and State timber, we never reached our export volume limit. We no longer exported Forest Service logs even though under the 1972 agreement we could have.

Each year, the Congress approves and passes to the President an appropriation bill. After 1972, it carried a stipulation that no monies could be appropriated for preparation of any National Forest timber sale for round log export west of the 180th parallel. Also, the Forest Service was directed by Congress to monitor purchasers of National Forest timber to ascertain they were not substituting National Forest timber for private timber under their control and allowing it to be exported in round log form.

However, people exporting private timber and operating mills were allowed to purchase National Forest timber for their mills from a third party. This pretty much negated the Congressional ban of National Forest timber from export and resulted in some overlapping regulations. No regulation could be written that would take in all the varied factors of the industry and the various types and species of timber. However, the basic intent of the restriction worked out quite well.

In our personal lives, Marzell, Jennie and I were shocked and saddened by the sudden death of Horace Alwood in 1974. He died in the shop of a heart attack. One of the drivers found him in the early morning, lying on the floor of the shop washroom.

Horace was a fine man of a strong Christian upbringing. Before World War II, he worked for the Turners in their service station and store on the Wishkah Road just below the bluff where the road is close to the river. He went to work for us as truck driver around 1939. Horace and DeLores Kerr were married in 1940 and after the war lived in Hoquiam. They had two children, a son and daughter.

After the World War II broke out, he volunteered into the Air Force and became part of the Signal Corps in the South Pacific. Horace was Canadian by birth. His parents at one time had a farm on the prairies of Alberta. Upon enlistment in the Air Corps he was granted full U.S. citizenship.

After less than two years in service he contracted a severe case of malaria and in 1944 received a medical discharge, came back home and went to work for us again as truck driver. Horace took medication for his condition caused by malaria.

As the company expanded, Marzell put Horace in full charge of all trucks, log exportation and shop management. Horace was responsible for much of the design of our shop and building put up in 1950, the location we are operating from today.

Horace always took charge of a situation. He never had to be told. When loading a vessel with logs or lumber he saw that correct documents were in place and that the super-cargo signed acceptance as agent for the buyer. The super-cargo was the agent aboard ship responsible to the purchasers of the cargo.

In 1972, Joe Ness and I were involved in the Milwaukee Land Co. purchase of 4,500 acres located in Elk River and Charlie Creek, just off the beach highway below Newskah, on Johns River, and also a section on the divide between the west fork and east fork of the Humptulips.

The purchase contained 35 million board feet of second-growth with a few acres of old-growth cedar in the Elk River area. We negotiated the purchase in the old White-Henry-Stewart building in Seattle. Mr. Bill Sanderson was land manager for Milwaukee Land Co. The total purchase price was $3.5 million, about $100/M. Land was worth about $60/acre. Most of the area contained a fine cover of young seedling and sapling hemlock. All the land had easy access and was fairly level. It certainly was an excellent addition to our tree farm.

Our total fee land base was now over 30,000 acres. It would have been a good time to quit, and sell out or merge. I don't know why we never got serious about doing something about building a stronger financial base. Probably it was because it was so much easier than our beginning years of the '30s and '40s.

We were always chafing to grab off another big project and it came in the form of the two Rayonier cedar purchases on Joe Creek near Aloha, Washington.

The first purchase we made jointly with Nichimen Co. It was about 35 million board feet at $80/M. It was medium-sized cedar, with not much No. 1 but lots of nice No. 2 and 3 export logs with intermingled 120-year-old hemlock.

Access roads were near the stand, and with the purchase we obtained gravel for road use at the cost of pit development.

Mr. Ishii, a famous Nichimen log customer, came over from Wakayama, Japan, to see the cedar timber purchase. He was well impressed by the straight, round, clean trunks of the cedar trees.

Mr. Ishii asked if there were any black bears in that area. He wanted one for the children of Wakayama, Japan, and he would build a grotto for the bear if he could get one. After Mr. Ishii left, I asked the Nichimen people, "Is Mr. Ishii serious? Does he really want to obtain a Washington black bear for the zoo in Wakayama, Japan?"

The answer was, "Yes, Mr. Ishii is serious and wants a young black bear for the Wakayama Zoo."

"That being the case," I replied, "we will see that Mr. Ishii gets a black bear cub delivered by aircraft to Osaka, Japan, the nearest destination to Wakayama."

We found a spot where a cub bear had been working at a honey bee nest in a downed cedar. We got a cage from a wildlife trapper at Sequim. While the bear was off on a jaunt, trappers made a trap which led the bear into the cage with a good supply of honey. When the bear went in, "snap," the trap shut her in. What a mad, snarling bear we had. The boys at the shop where we hauled her to lifted the lid to get a better look and found she was not hospitable.

To make a long story short, Northwest Orient Airlines had a special animal cage suitable for airlifting animals used in zoos. We got a black bear on the airplane at Seattle and off to Osaka.

Some time later, the Nichimen agent, George Horii, told me what a time the Nichimen staff had getting the cage with the bear in it off the plane, through customs and quarantine and into a car that the cage fit in and finally down to Wakayama. Then, the grotto for the bear was not ready, so Mr. Ishii had his workers build a temporary home for the bear. They named her Josephine, after Joe Creek, site of the famous Red cedar forest.

About a month later, Jennie and I went to Japan and to Wakayama, population 350,000. We were invited to a luncheon at City Hall by the Lord Mayor at which there were many dignitaries of the City. We were given the key to the city, made honorary citizens of the city and given a fancy plaque which I can't read. Then we went to see how well the bear had arrived, and after visiting the bear we were invited to see Wakayama Castle, up flight after flight of stone steps.

Josephine the bear, bound for Japan from the Peninsula.

Ten years later the black bear was still doing well and was a great source of joy to the children of that city.

Jennie and I made seven trips to Japan in connection with our log business between 1966 and 1979. Often we were guests of the trading companies doing log business with our company. One of Pacific Lumber and Shipping Co.'s main log customers were Shin Ashigawa, at that period a very important importer of logs in Japan. During those years we traded mainly with them. However, occasionally we would sell a cargo or part cargo to one of the many other Japanese trading companies importing western softwood logs.

Love of fine woods is a long-time tradition with the Japanese people and most all housing, from single family to fourplexes, are wood-frame construction with clay tile roofs. In the countryside, best seen while riding on the Shin Kansen, are old, large, traditional houses with thatched

Grays Harbor lumber at work in Japan.

roofs. The Shin Kansen is the "Bullet Train" from Tokyo to Osaka. It now goes clear to the southern island of Kyusho on two parallel tracks, one running north, the other south, often times elevated, at 180 to 200 KM per hour

We were fortunate to view the 1970 World Fair at Osaka, receiving V.I.P. transportation to the fair ground area through Pacific Lumber & Shipping and their Tokyo office. We have never seen another world fair, except for a glimpse of the Seattle Fair.

In the middle 1970s, the oil crunch seemed to have fizzled out, and the demand for lumber and logs seemed to get stronger year by year.

Len Forrest, manager of Rayonier, called and mentioned they had another cedar timber sale ready to sell. He asked if we had any interest. The Nichimen people were in our offices when he called, inquiring about another sale, but for some reason we did not know, the Nichimen people could not get permission from their Tokyo office to join with us in this purchase.

After the Nichimen people left our office, I called Mr. Forrest and said we would buy the timber ourselves and to proceed with necessary documents on the same terms and conditions of the first sale.

This sale contained 30 million board feet of which 60 percent was good quality old-growth Red cedar and the balance 120-year-old hemlock. Some of the cedar were large and contained more No. 1 grade. It was a very fine purchase.

Later, Rayonier had several more large timber sales in the area, but Mitsui Co. and Nissho Iwai entered the bidding and the price was more than double what we had paid.

At this time, Evans Products people became concerned with Indian tribal activism and their exposure on a contract they had on Taholah Unit. Individuals outside the tribe were raising a ruckus along roads and bridges outside of B.I.A. influence.

Evans approached us to explore our interest, if any, in the contract. For the contract, they wanted dollar for dollar of timber deposits, securities posted as bonds and felled timber at their cost plus $500,000 for "blue sky," the built-in profit in the basic contract.

Under Tom Mayr, our forestry staff, with Don Hurd, who had been with Evans Products, Gary Leonard and Mike Lentz, made a complete analysis of all the remaining cutting units, condition and volume of the felled timber, condition of the roads, and an estimate of the cost to bring the Chow-Chow Bridge up to safe standard.

At a regularly called Director's meeting, the Directors instructed me to proceed with negotiations for the Taholah Contract with the Evans Products people. We had no fear or concern that we could not continue a good relationship with the Quinault tribal people and make money on the B.I.A. Taholah Contract as offered to us.

Evans Products wished to have first right of refusal on any of the peelable cedar logs from the Taholah Contract by meeting the export log price. This was acceptable with us. It guaranteed us a market for the medium sized No. 3 cedar logs. Evans bought these peelers for their plywood plant in Aberdeen.

We met in the upper room of their Harbor Mill office. We were represented by our attorney, Mr. Clark Adams, Marzell, Tom and myself; Evans by Dave Davis, Peter Koehrler and Carl V. Egge, attorney. After several sessions we hammered out an agreement, which was approved later by the B.I.A. There were no considerations for mill, equipment, or labor contracts, just the Taholah Contract, dollar for dollar, plus the "blue sky". I thought it was a great day for us. We, both contractors and company, were very soon in full operation with 10 logging sides. Tom Mayr and Don Hurd and staff looked after the whole operation.

The market for Red cedar logs remained strong during most of those years. However, toward the end of the contract, there developed over 70 percent low-quality shingle cedar which could not be absorbed into the market and was put into inventory. By end of the contract we had 10 million board feet of shingle logs located at a storage area near Aloha Mill, under the bridge at Northwest Mill site and in the cedar yard east of the chip mill road.

The finest timber had been logged out of the unit many years before when Paul Smith and Aloha Lumber Company had the contract and were sawing the best cedar logs in their own mill. Paul Smith also operated two large shingle mills, one at Moclips, where he made a log pond by damming up the Moclips River. The other mill was at Lake Pleasant in Clallam County.

One time, Joe Ness and I were looking at some of the Polson property and the Moclips River was running good. Large salmon were trying to jump up on the dam apron to get past the dam, but they could never make it and fell back down and when they became completely exhausted, they died right there. An anadromus fish will only spawn in the same river or stream of its origin. What a waste.

Like other dams on other streams in the Northwest, the dam had no fish ladder. Nobody gave a good God damn about the anadromus fish resource so now the fish are all gone, and people are fighting over who gets the last few.

After we got fined for falling a few trees into the Miller Creek tributary of Clearwater River, and subsequently were charged by the State for their estimate of numbers of fish lost because of trees falling in the creek, we, especially myself, in complete ignorance, tried to do something about the anadromus fish.

On the middle fork of the Wishkah about 14 miles north of Aberdeen we owned 160 acres that straddled a beautiful stretch of the river. At considerable expense, we built a fish rearing facility. With the help of the State Department of Game, over ten years, we reared over a million steelhead, besides other fishes. We must have, in out-of-pocket cash, put close to a million dollars in that project.

Marzell never really approved it but I was blindly stubborn. Only now, after fifteen years, do I know how foolish a project it was. We really never got any cooperation from either State Department of Game or State Department of Fisheries.

Finally, we did get some help from the Quinault Indian Nation. They were allocating funds for the Chehalis Indian Fishery which was part of their fishing grounds by treaty. However, very few fish ever came back to the spillway; less than one fourth of one percent, I believe. If we had put that money in long-term treasury notes, we could have made gifts to the Salvation Army and the Union Gospel Mission at Thanksgiving and Christmas all the rest of our lives.

Our son Daniel was working with the rigging crew under Truman Santiago, an A-1, first-class logger. David was still going to high school.

It was the mid-1970s, and up to then, everyone had followed Marzell's and my lead. After all those years, we were thinking about continuity and the future. But, I could see things were beginning to follow a pattern and some day there would be a change, but I did not know how, what or when.

Bill Eastman, consulting forester from Seattle, contacted me saying Mrs. Arnold Polson and daughters wished to sell the balance of their Quinault Reservation holdings. Tom Mayr and I went to Seattle to meet Mrs. Polson. We were introduced to Mrs. Polson by Bill Eastman at her home, and got the necessary information.

Not too long before this, I had a casual meeting with my good friend Jim Jackson and he told me his dad Cleve Jackson was a sort of agent for Arnold Polson on Quinault Reservation matters while living. Whenever property or fraction of property became "fee patent" Cleve would secure it for Arnold Polson. This usually happened when a native having a trust allotment and married to a non-Indian died. The portion going to the non-Indian heir would pass directly in fee simple to the surviving spouse. However, if there were children, that portion passed to them in trust. Cleve Jackson helped Arnold Polson put together quite a few thousand acres on the Reservation in this manner.

When Arnold Polson and Cleve Jackson were gone, the Polsons asked Mrs. Jackson, Jim's mother, if she would sign a Quit Claim Deed in favor of the Polson estate. She gladly accommodated them but, according to Jim, they didn't give her one red cent and from that came the bitterness Jim held in the Polson deal. At best, he hoped some day the tribe would get the property.

The acreage we obtained in this purchase was a fee total of about 800 acres. This was the final purchase from the Polson family. We put quite a few millions of dollars into their pockets.

We had until March, 1979, to finish the Taholah contract. The acquisition of the unit and our involvement made it necessary to increase our management staff so we placed Clyde Stevenson at head of personnel. He was a gung-ho worker. Also, we brought in Urho Wirkkala of the famous Wirkkala logging family from Naselle. He was a very qualified road builder and was great help to Marzell who, for years, had built all the roads. We also brought in Jerry Logue as trucking supervisor. He came up from Madras, Oregon, having worked in a large pine mill operation and was experienced in all facets of trucking and log yard operations. In the mid 1970s our payroll roster was 450 persons. We were going full speed ahead.

Our comptroller for 25 years, Evar Carlson, retired, and it took some time to find a person that could step into his shoes. First, we brought in Matt Kuran, who had worked for Twin Harbors Lumber Co. in northern California. It finally became too much for him and we looked for a person to help him. We brought in Zane Keffel, a young fellow who had an education but little log and lumber experience. He was from Vancouver, Washington, and had worked for Dant & Russel in Portland, Oregon.

In those years of large volumes of export logs, mainly going to Japan, there were ship parties. For one aboard the motor ship "Asia Brightness," a vessel lifting a cargo of our logs for Nichimen customers, we compiled a list of 30 people from our supervisor's and office staff plus some of our close friends and business associates.

A ship party usually was hosted by the trading company and was held in the main dining room of the vessel, and began with cocktails and hors d'oeuvres. The ship's crew were at tables all dressed in their finest. The worst part, especially if the ship was light at high tide, was getting people, especially the ladies, through the long climb up the "gang plank." Of course, it had guard rails and also a net underneath.

The ship party usually included a full-course Japanese or Chinese dinner, and lasted about three hours, from 7 until 10 p.m. The longshoremen kept on loading cargo and probably let the logs bang around more than usual, putting on a little show.

One ship which came to the Port Dock in the early 1970s to load our cargo was the "Eastern Take." "Take" means bamboo in Chinese. She was only about 15,000 tons and her captain was an Englishman, the crew was Chinese. She was Hong Kong registry. This captain entertained our group several times when he came to pick up our logs. The vessel was on charter to Nichimen Co.

Another vessel I recall, the "Verdala," was of Norwegian registry. The captain spoke fluent English and invited Dan, Mike and I to have dinner with him aboard ship on a weekend. He was the nicest person you would ever expect to meet.

In my memory, Dad used to say "We will do it when my ship comes in." Ships came in often for us. If our dad could have seen us he would have been very proud of his sons. What would have made Dad happy most of all was our integrity, our respect for each other and lack of any greed or jealousy on the part of either of us.

The "Eastern Take" loading western hemlock at the Port Dock in Aberdeen. She was the scene of some of the ship parties of the Seventies.
Photo by Jones Photo Co.

That came later, when the children became adults. Our long-time attorney and confidant, Clark Adams, warned me about this long ago.

There were still timber deals out there waiting to be snapped up. Our good friend Maxwell Carlson had a section down near the wilderness which he and his brother Lawrence inherited from their dad, Gus Carlson. It had been a nice stand of 80-year-old hemlock. Art Shelgreen had logged about two-thirds of it and we cleaned up the remaining timber in 1943.

It was a coincidence that we had a chance to buy the section from Max Carlson, and we paid $850,000 cash. After all that logging it still had three to four million board feet of merchantable hemlock. In my estimation, that was some of the finest timber-growing land in the whole country and the ground was easy; rolling, well drained and no swamps that amounted to much.

Nissho Iwai had sale of a block of cedar and hemlock located near Clallam Bay on the Crown Zellerbach road system. It included about 50 million board feet, of which two-thirds was big swell-butted cedar and the rest mixed stands of second-growth and old-growth hemlock. We agreed to pay $10 million for the entire property.

Checks were made out to Mr. Morimoto of Federal Way. He had been local manager of Nissho in the 1960s and early '70s. I think he was part of the Nissho group, that Nissho really owned the timber and some of the people had some kind of funny profit sharing venture going.

Simultaneously with the timber purchase contract we received a very lucrative log sale agreement with Nissho at a price that we could make $100/M on all the export logs. We logged out 25 million feet the first year. In the end, we dearly paid when they quit buying the logs off the tract.

Then, the cedar market went all to hell and we had five million board feet of cedar decked at the Old West Coast Plywood yard. Hemlock, especially Clallam Bay quality, was also very weak. Nissho asked us to shut the operation down and we did.

Then, Mr. Morimoto wanted the rest of his money. We negotiated a cash-out price considerably less than the $5 million remaining on the contracts. I think we paid out about $2.5 million and received clear title, but we made the purchase with borrowed money. As it turned out, the market for this type of cedar remained in the doldrums for many years.

We were working steady on the Taholah contract. It had a March, 1978, expiration date, after which no more timber could be felled. However, upon felling, the timber became personal property and we had one extra year to remove the felled logs.

The Quinault tribe wanted the Chow-Chow Bridge left intact. It was in good condition so they and the B.I.A. had no concern about its condition. The roads only needed to be graded and the wasted ballast pulled out of the ditches and put back on the roads after the logs were cleaned up and all stumpage paid to qualify for a complete release, and we did obtain that release.

Toward the end of the 1970s, stumpage prices got so high that they surpassed the log value. It was no longer possible to buy one month, then turn around in two months, liquidate the timber, make a good profit and end up with the land and marginal corners as profit.

In 1978 we quit exporting private timber in our working area, and qualified for National Forest bids again. At that time the market was the worst it had been for many years. In 1979 things picked right up again, wilder than ever. It really was remarkable how in a few short months such drastic market differences could occur.

Two major timber sales came up. The Forest Service offered Cougar on September 30, 1977, 32 million board feet located on east fork of the Humptulips and upper Wishkah and, on March 3, 1978, a 14-million-board-foot timber sale on the Bogachiel, including nine miles of new road construction. Both sales attracted little interest, yet on State sales in the Clearwater, timber was going from $250 to $300/M with considerable road projects to reach the cutting units.

Rayonier was going along at merry clip in their new sawmill, using a surrogate bidder on Forest Service sales, usually Northwest Logging. In 1978 and 1979, timber on National Forest sales in the Grays Harbor Unit was going for over $200/M for white wood. Our company bid one small Forest Service sale on Salmon River, about three and a half million board feet. Bid price on white woods was $240/M.

The gyrations of the lumber and log market changed during the 1970s from earlier patterns. In a period of six months there could be a three month period on the downside and then three months hotter than hell, like ridges and valleys. This phenomenon affected bid prices on both State and Forest Services sales. Bidders were prone to use only the high market values and bridge over completely the low market values.

Many years before, we purchased the old Dillard farm at Vesta Creek on North River. It had around 80 acres of cleared pasture land, and I (I better not say "we," because any project that became wasteful should be credited to me.) planted all of the pasture with Douglas fir on the upland and cottonwood whips on the lowland, closer to the river. After fighting off the mice who chewed off the bark on the new whips and the water beaver that came up from the river, they finally did take off and closed in, making a nice stand of cottonwood trees.

In the final analysis, the only market for cottonwood logs is for pulp chips. I did see cottonwood plywood face panels some years ago in the Anderson Middleton Lumber Co. office in Aberdeen, but there is not sufficient cottonwood available in this area to justify going into cottonwood veneer production and it is not permitted for construction lumber under the lumber grades and construction codes.

But, the trees smell like honey in April and May when the buds begin to burst. Only a person who grew up in an area where there are these species growing in the wild would know what I am writing about. That sweet smell is a sure sign summer is near, kind of like a blue grouse, hooting away off on some distant hill, probably sitting on a limb of some gnarly old hemlock tree. Springtime in our far western forests is something most people, unless they work in the forest, are just not able to enjoy.

I became acquainted with Brooks Ragen, an investment counselor whose office was in the Washington Building in Seattle. He was in charge of finding a market for the assets of Twin Harbors Lumber Co., owned by our friend Heine Anderson and family. They had a very efficient, high-speed double-cut band mill. The entire mill was set up for eight-foot studs and, on good-sized logs, 20 to 30 inches, the mill could produce 200,000 board feet in one shift. Their Willapa operation had dry kilns, boilers, shipping shed and rail spur, and, above all, a first-class productive work force.

Their timber portfolio consisted of 7,000 acres of quality timber land, very well stocked with 80 million board feet of timber. The mill also had a large log inventory, around 6 million board feet.

Mr. Ragen wanted us to look over the property and secured a 30-day exclusive option from the owners. We checked with our bankers and lined up credit. The bank name was changed from National Bank of Commerce to Rainier Bank, and a Mr. Trueax was head officer, in the position Maxwell Carlson had for so many years. Mr. Trueax, of course, was a big shot. He didn't know us from anyone else in any one of the bank's many branches.

Tom Mayr, with his staff, looked over many of the cutting units on which there was merchantable timber, and we agreed to buy the specific capital assets of Twin Harbors for around $15 million cash, all with money borrowed from Rainier Bank.

After all those years we were still speculating on the market, but in the first year operation we were able to reduce the purchase price debt by 50 percent. Had we continued liquidating the purchased timber, which had a $100/M cost, we could have paid off the whole purchase price by end of 1980.

Instead, we started bidding the junky state sales in the Willapa area using the abnormally high 1979 prices. Some of the state sales we bought were "rehab" sales, with lots of brushy Alder and Vine Maple patches. We even purchased a low-quality second-growth sale, "Parpala," which was just south of the Naselle River and other sales near Salmon River east of Naselle. It was a nice truck haul into Raymond from the Naselle country.

We started operating the Raymond mill in January, 1979. Clyde Taylor, an excellent mill man, was in charge. Daniel was working with him, learning the sawmill business, and Mike was in charge of the Raymond log yard. We also black-topped the Raymond yard.

We were blind. Hell, all the timber of any size was gone in that area. All that was left was sapling-size hemlock and, here and there, a few patches of limby, second-growth fir. We had not yet logged any of our tree farm, except what old-growth there was on the timber purchases on the Reservation and Clearwater, Kalaloch and Hoh Rivers.

We had a continuing program of building roads into many of our advance second-growth stands on the Wishkah, Elk and Johns Rivers. Our tree farms were now over 40,000 acres located in six counties of western Washington, from Clallam Bay to the Columbia River.

About that time we heard from Sam Newman, a forester and land manager who was retained by the Kelly family for the purpose of cruising, appraising and looking for a buyer for two fine properties just above Hoodsport near Fulton Creek. The properties totalled about 280 acres which had been logged into Hoods Canal at least 50 years before.

It contained about 5 million board feet of mostly Douglas fir from six- to 20-inch butt diameter, a nice log for the our sawmill. There also was a small volume of cedar, hemlock and lowland White fir. Logging conditions were good for high lead. We bid $875,000 cash for the property and our bid was accepted.

The property had no access problems. Joe Ness took care of the operation and contracted the cutting, logging and trucking out to Don Makoveny who was an experienced logger and had access to some pretty fair equipment. His loggers were no fireballs, though, usually logging about 65 percent of normal production. On one corner of the property, north of Fulton Creek, a neighboring owner claimed we trespassed, cut some trees on his land, and the loggers hooked their tail blocks on some of his trees.

By re-establishing the property lines with a registered land surveyor, we found out we did not cut any trees on his side of the line. He still wanted damages for the trees we hung the haulback blocks, and we made good with him, arriving at a damage settlement. It did not amount to much, two small fir trees about 14 inches on the butt. The logs off this tract were fine grain and they enhanced our Douglas fir lumber sort.

I went up occasionally to observe the logging, and one of the crew pointed out to me that looking northeast you could see the nuclear submarine base across the Canal. It was, I guessed, about six miles away.

Timber bid prices during the last half of 1979 went completely out of hand. People were bidding to and over $400/M on 60 percent old-growth and 40 percent understory second-growth Clearwater hemlock. On a predominantly cedar sale on Mud Creek, called Mud Flats, Marzell bid against Jim Middleton of Anderson & Middleton. Our bid was $1,275/M. Jim bid once more $1,280/M, and they got the sale. That was at a time when the best No. 1 cedar was selling at around $900/M.

Like mentioned before, the bidding in most of the '70s, except for a slight slack period around the OPEC crisis and 1976-77, was purely speculative. The object was to get the timber sale and hope very soon the log value would rise by hundreds of dollars per thousand board feet.

I began to see the possibility that before too long there would be an acute shortage of old-growth Red cedar and possibly legislation passed on the Federal level prohibiting the export of unprocessed Red cedar logs, and that did materialize. U.S. Representative Don Bonker presented a bill in Congress restricting Red cedar round log exports which did pass.

On the Twin Harbors purchase we were logging full-tilt, and the market was good. By liquidating the timber into the round log export, we were allowed to export the Willapa area logs because they were out of our sawmill operating area and we were not bidding any National Forest for the Raymond mill.

In part of the tract known as the Oxbow block, there remained four settings of Douglas fir, from 120 to 180 years old, mixed with some very nice hemlock and a few large Sitka spruce.

Bob Burbank was logging supervisor on the high-lead side and he was getting good production. By early summer, we were getting near the valley water system watershed. Some of the logging we wished to do threatened to silt up their water supply. The creek had a major fork about a half mile above their intake dam, so we agreed to lay a six-inch diameter water line from the fork of the creek that would not be impacted by the logging and put a valve hook-up at their intake dam so the water could be shut off if any silt developed. The clear water from the stream that was not impacted by the logging fed into the system.

The valley water users were well satisfied, and we never heard any complaints. If they take care of their water line, sometime in the future whenever logging needs to be done on the unimpacted fork they can move the pipe over to the fork already logged. This is the same type of system used many years ago by the City of Aberdeen when the Wishkah watershed was logged.

During the fall of 1979, Brooks Ragen mentioned to me that he was in contact with a large company in the forest products business that wished to make a major expansion of their company base. It was an old, long-established company, Pacific Lumber of San Francisco, which had a huge redwood operation in northern California. They were operating an old-growth redwood mill in Scotia, California, and a new mill at Fortuna, California. This mill was cutting second-growth redwood logs up to 30 inches in diameter. It was a similar set up to our mill.

I talked to Marzell about this and he thought I was talking about Pacific Lumber & Shipping, so I told Brooks Ragen, "Certainly, we would like to talk with these people. Invite them up to talk."

In a couple of weeks, the Chairman of their Board, Tom Hoover, their President and Chief Executive Officer, Gene Elam, and head of manufacturing, Ted Malarky, came to Aberdeen from San Francisco to have a look-see at our operations, mills and timber land.

Brooks Ragen brought the visitors to Hoquiam for introduction and meetings. After cordial exchanges and pleasantries, I took them to examine our Hoquiam mill and the Raymond stud mill. On the next day I took them on an excursion of our timber-land holdings located on the upper Wishkah, which were well roaded and where no logging had yet been done.

Our completed fish rearing pond on the old Achey homestead was in operation. It was its second year and the pond had about 100,000 fingerling steelhead in it. The fish rearing pond is in a beautiful section of the upper valley. The place opens to the south with about 25 acres of meadow around the pond and all around are forested hills.

After two days looking at our operation, they headed back to Seattle and to San Francisco, and issued us an invitation to come down to Fortuna and Scotia to see their redwood operations.

After talking it over with Marzell, we decided Jennie and I should go down and look over their mills, log yards and forest operations. Several weeks later, Jennie and I caught a flight to San Francisco. We met Brooks Ragen in the San Francisco airport and flew north to Eureka in the same plane.

Eureka is a coastal city not much different than Aberdeen, though probably warmer and very foggy in the summer time. There are many old timber baron mansions in the city, dating back to the 1880s.

We met the Pacific Lumber people at Scotia. They took us by car out to their old-growth redwood operation where we met their chief forester, a very nice person. He told us the company had 10,000 acres of old-growth redwood running about 100,000 board feet to the acre plus 190,000 acres of second-growth redwood, much of it from 60 to 90 years old, and that they had just built a new mill at Fortuna which we would see on the next day.

We headed out through semi-ranching country not much different than the upper Wishkah and before their logging road broke into the foothills, we stopped and had a picnic in a little private park and observed some good things they were doing for game and fish habitat.

I had never been in a redwood operation. The logging was by arch and tractor and the steeper ground was being logged by short span high lead. The redwoods were four to seven feet in diameter. Many of the trees had very few branches and looked a lot like the huge, partly-dead

Red cedars on the Matheny Mud Creek State Forest or the Nolan Creek flats.

The head forester had a special treat for us, and we were able to watch a set of their best cutters drop a huge redwood. I watched him mark it out. The grain even had a finer ring count than found in some of the oldest Red cedar in our country. There were, scattered among the Redwood, a sort of 150- to 180-year-old Douglas fir, nice trees 20 to 36 inches on the butt.

It was early fall and the ground was damp, but just right for logging with tractor. We were able to watch them log a few turns and load some trucks. They had a private road system which, in the lower section, had been a railroad operation.

Pacific Lumber was founded around 1880 as a private company by the Murphy family and many years later it became a closely held public company. Its stock was traded "over the counter." It was evident to Jennie and I that it was a company of very high standards and tremendous wealth.

Late in the afternoon, we came back into Scotia, which at one time was a company town. Now, people owned their homes and the town of Scotia became incorporated at an earlier time. It had a large company store where you could buy almost anything. The town had a little park where there was a Washington steam yarder with huge drums. It was a long-span skyline machine from the railroad logging days.

The company had a guest house, a large, old-time mansion where corporate people stayed when they came up from corporate headquarters in San Francisco. This is where we stayed for the night. That evening we were served a scrumptious meal, and after dinner Brooks Ragen met with the Pacific Lumber Company executives and discussed whatever over cocktails. Jennie and I had our own king-size private room.

Next morning we had a very nice breakfast at the mansion and got everything packed. We then met some mill people who escorted us through their huge old redwood lumber mill.

The stringy redwood bark was a special barking problem. It came off in long strips. From there it went through something like a huge set of clothes wringers, and these squeezed a lot of the water out of the bark. From there it was chopped and hammered to be fed into one of four huge furnaces which heated huge boilers for steam to the turbine which turned generators to run the mill. After going through the turbine, the steam went to dry kilns.

Redwood lumber in the rough is "stuck" with spacers and piled outside to begin a slow process of seasoning, as long as one year, after which it is put in the steam-heated dry kiln. I don't remember the number of hours it is in the kiln, but I imagine at least 12 days because old-growth redwood is a very heavy, water-soaked wood and when such fine-grain, high-quality wood comes out it must be dried to absolute perfect moisture content.

Much of this high-grade finish paneling ten years ago sold for $1,800/M in six-, eight- and ten-inch widths. Panel boards, some up to 18 inches wide, would bring twice that price.

We went into their warehouse and shipping shed. I have no idea how much finished lumber was in the warehouse. Someone said the inventory in lumber was $40 million, and it was all their own money.

Our motor caravan headed for Fortuna, where we observed their new mill, similar to ours, which cut second-growth redwood logs yielding knotty one-by-six and one-by-eight paneling and all sizes of dimension lumber.

In a later conversation with Brooks Ragen, he told me that from what they had seen looking over our operations, they were very interested, and we could have an exchange of some cash and receive shares of Pacific Lumber in exchange for Mayr shares. According to Mr. Ragen, they very much desired a company with healthy growing resources. If such a merger could be worked out, there would be at least one seat on their board of directors for the holder of our shares.

Apparently, Pacific Lumber's timber portfolio was generating tremendous profit from harvest of their low-cost redwood timber stands. Of course, a sale of their assets would never bring them anywhere near the true value of their holdings so it was best planning to merge and expand.

After lunch, the Pacific Lumber people took us back to Eureka airport and Jennie and I flew back to San Francisco. We had accommodations at the St. Francis Hotel, and the next morning, we went by cab to San Francisco airport to catch our flight back to Seattle, where we had our car parked in a airport parking lot which was served every few minutes by private bus. We arrived home to our family at late evening of the fourth day. Cathy and Oley were looking after the children.

I asked Marzell for a conference. We sat down and I related our trip to San Francisco and the Pacific Lumber operations in the forest and the saw mills. Apparently I did not do a very good job of telling Marzell what an opportunity it might be for us. He asked if it was part of Pacific Lumber and Shipping, located in Seattle, but did not express any further interest. We told Brooks Ragen that we had no interest in such a venture at present time.

Hemlock for export at the yard in North Hoquiam. These logs would be shipped after the parade for Logger's Playday.

Our company had a good financial report for the first three quarters of 1979 and came out at year end with $4 million profit. We had log export market connections which we thought would enable the company to purchase State, County and Indian timber for the export market which would not adversely affect our bidding qualification to purchase National Forest timber for our sawmill. 1979 was a peak year in volume for the company in log and lumber production sales.

The last months of the 1970s were hectic times. Our operation had expanded tremendously. Our total employment was approaching 500 people, and the two sawmills and the chip mill were consuming 4.5 million board feet of logs per month.

We were selling into the export market 5 million board feet per month originating from State and Indian timber sales, and selling cedar logs into the shake and shingle market. Our working log inventory was about 25 million board feet, of which 20 million was logs held for resale into export or domestic market.

After the successful purchase of the Twin Harbors lumber operation we were judged by the bankers as people that rarely made bad deals, at least up to that time. We had established with Rainier Bank a $25 million credit line, divided into categories; banker's acceptances, $6 million; equipment, $3 million; mill, $3 million; timber lands, $10 million; and inventory, $3 million.

The banker's acceptance amounts would be retotaled as export logs were loaded aboard ship and payment received. New cargo upon which firm order was placed would qualify for new borrowings. Equipment loans were for use in buying mill and logging equipment or trucks. The mill was security on operating line, and the timber lands portion of $10 million would be repaid when timber land was pledged to Federal Land Bank on long term loan. The inventory loan of $3 million also was a rotating loan.

Interest on all loans was paid monthly. Interest on bankers acceptance was at 30- to 60-days banker's acceptance market. Timber land loans were prime plus one-and-a-half percent, and the balance was prime plus two percent.

The amount borrowed at one time never reached the full amount of the loan commitment and after placing the timber land under the Federal Land Bank the commitment by Rainier Bank dropped to $15 million.

Rainier Bank of Aberdeen continued as the main bank for our short-term needs. After changing from National Bank of Commerce, it set up commercial lending departments. Southwest Washington was served by the Tacoma banking center, and Ron Jutila was manager and the person with whom we dealt. However, loans of the size we used were considered by the senior loan committee in Seattle.

After the change in the bank's loan strategy, we became further removed from the bank's decision makers. Our friend Ron Jutila became more of a messenger, presenting a borrower's needs to the senior loan committee, where we were not present, but that seemed to work well. It did permit much easier "closing of the gate" should they decide a change was warranted, and that was to manifest itself very graphically in the change of their attitude to the forest products industry on a down-side market.

Our dealings with the Federal Land Bank of Chehalis, a branch of the Federal Land Bank of Spokane, began while they had a local office in Elma under management of Mr. Don Whittaker. Our first loan from the Land Bank was closed in 1960 for $550,000, at five-and-a-half percent. It was a 30 year loan for which we pledged about 4,000 acres of the earliest timber and land we had purchased. Interest and principal payments were made annually. By the late 1970s we had borrowed from the Land Bank about $20 million secured by a pledge of 20,000 acres including 250 million board feet of merchantable timber. The bank was setting a value of $80 to $90/M to the timber value which was then very conservative.

We were reaching our peak production and had built in far too much logging capacity in light of the availability of timber supply in the Grays and Willapa Harbors area. At that time the only remaining timber supply was on the National Forests on the west side of the Olympics. We were already the largest buyer of State timber sales in the Queets-Clearwater area, plus we were buying sales in Pacific County and in the Capital Forest.

One remaining timber area was the Makah Reservation in far northwestern Clallam County. It was a 140-mile truck haul to the Harbor and about 60 miles to Port Angeles. The Makah tribe business committee chairman, along with several other tribal members, came to our office during 1979 to discuss the possibility, or interest on our part, of participating in harvest of the remaining old-growth timber on their reservation. They had offered several timber sales during a high market period but had received no bids. Their mission at our place was an attempt to interest us in buying some of their timber sales.

We said we were looking for additional timber supply. Before long, we would be finished with Clallam Bay timber purchased from the Key Corporation. We discussed the quality of their timber.

It was all near the coast on the far northwest corner of the Olympic Peninsula. Continually buffeted by storms, the timber on northern and western slopes was rough of bole, with erratic grain, short trunks, and very heavy limbs. It usually would only be accepted by the export market during periods of high demand that coincided with periods of manpower shortages making it difficult to get production. On a poor market, a very high percentage of the timber would only be suitable for pulp chips.

We told them that the only way we could come into the Neah Bay reservation timber was on the full understanding that if we could not make money on account of market conditions, we would be permitted to shut down without penalty and extension of time to remove timber without penalty would be granted to us.

They said they understood that, and we agreed to study their timber resources.

In December, 1979, because it had been a good year, Jennie, as secretary of the corporation, and myself, as president, hosted a Christmas party at the Nordic Restaurant in Aberdeen. All our supervisory people and our log and lumber customers were invited. Many of our suppliers and trade people were invited, even people from the U.S. Forest Service. It was a gala affair.

We had Christmas trees, and even an ice carving. It was really a fitting thank-you, but little did we know, or even dream, that it was the absolute peak of Marzell's, Jennie's and my business career.

We had a very large log inventory, almost 10 million board feet of shingle cedar, plus 15 million board feet of export logs plus the domestic sawmill inventory. Our relations with both the Federal Land Bank and Rainier Bank were at optimum peak. But, interest rates were going up and an election was coming in 1980 and we didn't know what kind of administration would follow President Carter.

Everyone had a good time at the party. That night or next day, even the Christmas trees disappeared. Neither Marzell nor I got one for ourselves.

Chapter 8: The Eighties

The Sky Caves In

DURING EARLY 1980, MARKET DEMAND began to slow down, immediately affecting log and lumber prices, and the cost of money rose drastically. From 1978 to mid-1979, the prime rate hovered between nine-and-a-quarter to 11.75 percent, but by January, 1980, the prime rate reached 20.5 percent.

Our business was an enormous user of borrowed capital, and our inventory exceeded 30 million board feet of logs and lumber at an average cost of $300/M for a total of $9 million in loans just for product inventory. This inventory was three times normal, and it burdened us through all of 1980 and until June, 1981, when it was liquidated at an average of $100/M below cost. Between the interest and value loss, in the one year we lost $12 million, all of the equity we had built into the company over a period of 40 years.

I felt the market for logs and lumber would turn around. It always had since the 1930s, but this time it took much longer. However, we had excellent bank credit with both the Federal Land Bank and Rainier Bank. They felt secure, so we felt secure.

The bidding on public timber sales continued to be strong and aggressive. State of Washington timber for export was still going at $300/M for hemlock and $350 to $400/M for 100-year-old Douglas fir. This strength at the bidding floor on both exportable and restricted for domestic use gave me a false sense of security.

As part of our timber supply planning, we decided to examine timber on the Makah Indian Reservation. The Makah people still had 80 million board feet of old-growth hemlock, some low-grade cedar and medium-grade Sitka spruce. Don Hurd, Tom Mayr, and

Gary Leonard spent quite a few days on the Makah Reservation cruising and evaluating marketable value of the timber. Although the Makah area timber was sound, the grade was horrible for old-growth timber, 95 per cent No. 2 and 3 and the Red cedar was worse.

On the nicest round No. 2 and 3 cedar, the natives stripped off the bark while the tree was standing. They cut out a wedge of bark at the base of the tree with an axe, loosened the bark on two edges until a strip four to six inches wide was free. Then they would grab it with both hands and, pulling and jerking, pull the strip of bark off the tree, then repeat the same process until the whole tree was stripped of bark, 20 to 30 feet up on a clear tree. Of course, the tree died then.

This could only be done when the sap was running. This affected the value of the tree for lumber because sap stain would set into the wet sap wood. The bark was used for basket weaving and floor mats.

The B.I.A. reappraised the Makah timber sales downward, and readvertised them, but there were still some severe requirements in their prospectus, including removal of any log with ten board feet or more, even Red alder. Still, it looked workable, assuming the market came back. The buyer would have 32 million board feet of exportable timber available. We also assumed we would be able to work with the Makah like we were able to work with the Quinaults. We found later, to our complete surprise, they were altogether different. They were after money, and all they could get.

We put minimum bids on the Bear Creek No. 2 and Cape Flattery No. 2 sales in September, 1980. We were the only bidder and were awarded the sales. After posting the $400,000 performance bond and advance cutting deposits, we started logging with two contractors, Morrison's Clearwater Logging Co. and Jack Adams Logging Co., hauling the logs to our Hoquiam log yard.

Both loggers went broke up there. Marzell said anybody that tried buying or logging Makah timber at Neah Bay went broke. He was sure right.

We took over two smaller timber sales in the same area. The first was Jumpoff, from Hyusing America, a Korean company. Their subsidiary, Unipac, had a sale with high-priced stumpage on the Makah reservation on which the timber was felled and bucked.

The contract was due to expire, and the B.I.A. was going to foreclose. Unipac asked us if we wanted the contract, and we agreed to take over if they would purchase a cargo of logs at $100/M over market to cover the loss on cleaning up the contract.

We logged and completed the Jumpoff sale, but Hyusing didn't live up to the agreement. Instead of paying us $100 over the contract price, they paid us $100 under the contract price. We lost $700,000.

The Blowdown '79 sale we took over from a logger named Gardener. It had some nice exportable hemlock and spruce and at stumpage prices would pay out at a profit.

Practically from the start, we lost money on all logging in the Neah Bay area of the Makah sale. Usually the log market would firm up during the winter and put us in a profit mode. Instead, the market continued to drop. Also, log quality standards were getting more rigid all the time and the Neah Bay timber was, at best, not appealing to the overseas log buyers from Japan and Korea.

We were beginning to log on our tree farm, assuming cheaper stumpage and good-quality hemlock and second-growth fir would offset the losses on State and Makah contracts, but after factoring in the high cost of money, there were only one or two months in the years 1980 through 1982 that we were able to make even a small profit.

Our mills were still making a profit, if interest cost was not factored in, but at the Raymond stud mill, month by month, log quality and size became steadily worse. There were too many small logs and too many short logs. We struggled through 1980-'81 steadily losing money. After the purchase of the 35 million board feet of Makah timber and the continuing losses, I feel our main bank, Rainier, got concerned but never cared to sit down and help us put the high-priced sales on a minimum-cost hold pattern.

Ronald Reagan took office in January, 1981. The new administration's avowed purpose was to stop inflation by increasing the cost of money and strengthen the dollar, which adversely impacted the export market for both logs and lumber. Our offshore trading partners, with their cheaper currency, were hard pressed to buy U.S. products, which in turn put a throttle on our export market.

The growing strength of U.S. currency had an immediate adverse effect on U.S. pulp and paper production versus Scandanavian and, to a lesser extent, Canadian production. Efficiency of production in foreign pulp and paper mills was head and shoulders above U.S. mills at the market place for pulp and paper. U.S. currency continued to strengthen, and U.S. pulp and paper production stayed in the doldrums for many years, until foreign currency began to strengthen in 1985-'86.

We did, in 1983, receive some contract modifications on the Makah Indian contracts at Neah Bay but the market for Neah Bay type hemlock was about equal to the stumpage being paid. Good quality hemlock dimension-type logs were $160 to $180/M for domestic processing, and select hemlock, average length 39 feet and average volume per piece, 400 board feet, was going F.O.B. Port Dock for $210/M, seller pays Port charges.

The so-called knowledgeable people wanted to blame our strong currency on the huge national debt, but if the nations that owed the United States began to pay back money instead of borrowing more, our national debt problem would be vastly different. The United States government is not a good money manager.

Our remaining hope on the Makah timber sale was to salvage as much of our investment as we could, if we only recovered fifty cents on the dollar.

In 1983 we received a letter from Ron Julita stating the bank would not finance any of our company losses on the Makah timber sales. Both contractors moved their equipment out. Later in the fall we put our Washington 208 slackline yarder to work with our own crew to further salvage what we could of our investment in prepaid stumpage and felled timber.

Marzell, Dale Colton, and David Mayr were up on the job looking after the operation. Truman Santiago was hook tender. Also, we had the 041 Madill yarding the second felled cutting unit on which Raleigh Frank was hook tender.

We worked on sale area improvements until the B.I.A. ordered us to cease operations, after which we moved all our equipment off the Makah Reservation. It was a sad day for the company. This was the first time in our history we were unable to complete a timber contract. I am sure the $400,000 bond on the contract is the factor that affected their decision to throw us off the Makah contracts.

The fall of 1983 was the fiftieth year in business for Marzell, Jennie, and I, and we had a real nice picnic up at the Achey place, inviting many of our friends, business associates, and people working with us. It was a nice fall day, there was plenty to eat and people appeared to have a good time.

I don't think I have even been up to the Achey place since that time. I remember when we bought it from Mort Achey for $11,000. One time we even fixed the fences, built a new barn, and generally tried to keep the place up and finally built the fish pond referred to earlier.

During the fall of 1983 and winter of 1984, our main bank, Rainier, in the persons of Mr. Julita and Mr. Ingram, was calling on us once a week, demanding lots of paperwork, spread sheets showing cash flow, cost, and cost control.

The bank put us on a "BanControl" account under which all checks and receipts were sent direct to the bank by our customers. After a one-day float, the bank applied funds received to reduce our debt to them. In order to meet daily cash needs for payroll, taxes, stumpage, etc., Don Deschenses, our office manager, would let the bank know daily what our cash needs were and they would transfer that amount to our account so the checks would be honored. That was a galling experience for me.

Following on Mr. Julita and Mr. Ingram's visit to our office, we received word from them to reduce our loan amount by $1 million per month. This was impossible because the total inventory value had fallen far below its cost. They did not even seem to care. They were following instructions given to them.

Further pain and embarrassment followed. We were put under Mr. Craig Fitters of "Special Credits." The noose was tightening. If we could have gotten Rainier Bank paid off by liquidation of assets, there would have been a chance to survive.

The cost of money remained very high. As a matter of fact, prime rate remained higher than anyone had seen it for the third year in a row.

The Federal Land Bank of Spokane looked like a white knight coming to save the company by moving the remaining timber debt from Rainier Bank, issuing us a final consolidation timber loan in mid-1981 when money cost was already climbing. We were able to save interest cost of 11 percent by moving from Rainier at 22 percent to Federal Land Bank rate of 11 percent. During this move, Rainier Bank people alluded that they would never let us down and were deeply committed to the forest products industry and they would continue to support us. This, of course, made me feel good, at least for a few days.

The market for our products, export logs, lumber from the two mills, and pulp chips, continued to sink lower and lower. The higher interest rates went, the stronger the dollar became, and the lower our product prices became. President Reagan's battle for control of inflation was working well.

Contrary to the belief of Marzell that we needed improved efficiency, I think we had good cost control and good production in most all areas. Marzell was right in that an operation needs efficiency at all times, but the most inefficient areas were improper bucking of all the low grade timber we were working in, and all the pulp and cull logs coming in with prime production. This severely impacted log yard costs.

There was no way we could make anything out of the poor stands coming from Willapa Harbor rehab sales, Iron Springs, and some of our private purchases. The stumpage on State sales precluded coming out at a profit.

In 1980 we purchased a 14-million-board-foot, 80-year-old Douglas fir timber sale from Port Blakely Co., along with several other sales, and we neglected to get firm, signed purchase orders from Nissho Iwai, which, at the time, we could have gotten. This resulted in additional losses when the log market dropped below cost. Nissho were unwilling or unable to hold the log price at the level which we used at the time timber was purchased.

Early in 1983, at the invitation of Nissho Iwai, Ron Vandiver, Chief Check Scaler and scaling teacher for the Grays Harbor Log Scaling and Grading Bureau, and I took a trip to mainland China on a mission to teach our method of scaling logs. When the Chinese measured a log, they dropped the fraction part of an inch, and our scalers averaged out fractions in diameter as was custom and part of scaler instructions.

China did not interest me too much. It is a huge country, but not at all like Japan or Korea. Their demand was also down, which affected the market for fir logs to China.

As 1983 wasted away, we kept blind hope that the market would firm up in the spring. Daniel was selling lumber and logs and Don Deschenses was doing his very best to keep bills paid up. I was always thankful he kept withholding and Social Security taxes paid up, and the stumpage payments were kept up.

Don Deschenses would tell Dan Mayr he needed "X" number of dollars to meet payroll, and Dan was usually successful in getting the money from Niisho or the Korean Co. under Peter Megurian. It got so tight, we asked for advance payment on decks of logs which were secured to Rainier Bank and made payment to the bank as logs went aboard ship. We told the bank it was necessary to "double dip" in this manner, because the bank would not put up any "new money." I will never forget Bob Ingram, III, telling us, "We, the bank, cannot loan money to you if you are going to lose it."

As spring and early summer came in 1984, Dan and Cathy's husband, Ole Mackey, were working together to get some cheap logs and sell lumber for the best prices available. Interest cost was eating us alive. Rainier Bank took a second mortgage on lands we pledged to Federal Land Bank of Spokane. The Land Bank took a first mortgage on the Raymond mill and mill site.

In June, 1984, word came from Craig Fetters, of Special Credits: "Liquidate your trucks, logging, and road building equipment and get out of the logging business."

We picked Ritchie Bros. auctioneers, specialists in liquidating trucks and logging equipment. All the equipment was brought in and prepared

August 9, 1984: "Liquidate your trucks, logging and road-building equipment, and get out of the logging business." Scene from the auction, below.

for auction.

Sale date was August 9, 1984. It was a sad day. Jennie and I, with our youngest boy, John, went to our place in Canada. John, then 15 years of age, liked to hunt and fish.

At the sale, most of the equipment brought low numbers. The Sparmatic, had it been new in 1984, would have cost at least $300,000, and even though it was 12 years old, that kind of machine does not lose its effectiveness. At auction, it brought $8,000. It could have been scrapped out and brought more than that. The sale brought in about one-third of the equipment's value in good used condition.

We got back from Canada a few days after the sale, and I received a call from Ron Julita, which surprised me. He asked if I would instruct

Ritchie Bros. to make the check for the equipment sale receipts to Rainier Bank.

"Yes," I said, "I would do that."

Maybe they felt that if we were forced into bankruptcy the court might judge they had received preferential treatment over other creditors, because they had forced us out of business and therefore may have harmed some of the creditors.

We continued operating the mill, but now the bank knew they were under-secured and additional funds advanced might be like throwing good money into a lost cause.

Time for a major decision had arrived. After consulting with Cressman, we decided to file Chapter 11, which we did on August 28, 1984. Rainier Bank, with control of the inventory and receivables, had us locked up tight and we had no private funds to operate the mill. We were going to Seattle negotiating with Craig Fetters on some plan to reopen the Hoquiam sawmill and chip mill, which we managed to do in November.

The bank brought in the firm of Management Technology Associates, with Mssrs. Hagga, Moser and Franklin, to decide whether to continue operations of the mills or just shut them down and sell off the operation, "cold turkey." The M.T.A. boys brought in a slug of laptop computers and started their study to determine feasibility.

They came up with a recommendation to the bank. The mill always made money and still could make money. The bank instructed them to draw up an operating plan, which they did.

To us, the bank said, "We want $75,000 a month protection payments, and the company must hire an outside manager. The company is to have weekly management meetings chaired by the manager and the bank will send a representative to those meetings. The company no longer has any operating capital, so the bank will provide $475,000 new capital secured by the shipping shed, inventory and receivables on an 'as-needed' basis." This amount was later increased to $750,000 with a maximum of $850,000 for winter inventory.

Other stipulations were that M.T.A. would be present at our weekly management meetings at a cost of $4,000 per week, plus expenses and family members in the company could only receive a maximum salary of $3,000 per month.

We did not see how, under those market conditions, the mill could support such an excessive overhead. The position of the bank and M.T.A. was diametrically opposed to how I felt, and even Marzell believed that way.

Tom Mayr, President of MB Logging, 1985.   Photo © John Tylczak

Their philosophy seemed to be, "When a horse is down, beat the hell out of him and put the fear of God in him and he will work like hell," the idea being that we were asleep at the switch before and that caused the problems in the first place. How could we have accomplished what we did if that were true?

Finally, we negotiated with the bank and signed all necessary papers and got the go-ahead to operate the mills at Hoquiam. Right after bankruptcy, our entire crew agreed to take a $2.05-per-hour pay cut and drop down to three paid holidays and only one week vacation with pay. If the company made any money above all expenses on a quarterly basis they would share and share alike out of 25 percent of those profits.

Our operation restarted November 1, 1984, on one shift. After

December, we went back on two shifts. We brought up the lumber and second-growth export reject logs from Raymond, plus what logs we had in the Hoquiam log yard.

Darrell Barnes was in the mill and I thought he handled it very well, but Marzell and his family didn't want him. We finally picked a professional sawmill manager, Marvin Faughnder, late of Simpson Lumber Co. He came in March, 1985, and did his best, but there just were no logs available of any quality at all.

We did get some old-growth spruce from Harold Brunstad, which originated on a Forest Service sale on the west fork of the Humptulips, and the price was right. We had the Elastic timber sale, on the east fork of the Humptulips. It was a high elevation sale.

The bank did not want the company involved in any logging, but they did permit us to petition the court for the right to set up MB Logging under Tom Mayr's sole control as a subsidiary of the parent Mayr Brother's Logging Co. Through transfer of some uncommitted assets, MB was able to get enough logging equipment to start one side during May of 1985, with a used Washington model 208 yarder.

The mill made a little money at first off the Brunstad logs, then incurred some huge losses. Marvin Faughnder resigned in October, 1985. There was no log inventory control and there were no logs available except export rejects and pulp logs from Anderson Middleton. Only Don Bell was logging a good quality old-growth sale at $13/M stumpage, but the logs were all committed to Dahlstrom Bros. The stumpage price was ludicrous.

Things were shaping up to what the future would be like if we were able to reorganize. Ole Mackey and Dan left the company. Also, the mill was getting old and cutting off size by as much as a quarter inch, which was costing recovery, and we had a struggle to get shop cuttings out of the logs. Don Deschenses, now manager, brought in Don Larson, an ex-Weyerhaeuser mill man versed in shop and cutting logs.

I never did think shop and high-grade cutting was his forte. He made a whole lot of paper reports, and at the next management meeting, I blew my top and said, "If we have to hire somebody to tell us how to get out the shop, then we need new people in the mill. A mill superintendent of a cutting mill must be a person that knows shop and clears or he should not be there. That is one thing about both Roy Aarhaus and Darrell Barnes, they knew lumber grades and a cutting mill's recovery aspects."

After the meeting, Marzell said I acted like I was drunk or on dope. I am sure D. Hagga knew what I was talking about, but he said I should not blow up. There was no way to lead, the family appeared to be taking positions and I didn't like it at all and could do nothing about it. Those were sleepless nights.

We brought in Dan Staley, Don Deschenses' uncle and an ex-savings and loan bank examiner, for the purpose of getting complete list of creditors and the company assets. I think he did a good job. It seemed to take such a long time, with over $40 million in debt to be sorted out, there were so many creditors.

The Federal Land Bank had personal guarantees from Ole, Cathy, Tom, Theresa, Marzell, Jennie and I, and everyone was afraid of losing their home which put us in zero position to negotiate. They said if we would sign all land secured to them, they would release our personal guarantees, but how could we continue in the mill business, in the long run, without timber and timber land? We were not well counseled at that point, and very shortly after, Federal Land Bank got title to our land. They sold it all off at, I guess, at forty cents on the dollar of debt.

All of our bonding needs since 1968 had been handled by Raleigh-Mann and Powell. They placed all of our insurance coverage and our bonding needs with United Pacific Co. and at bankruptcy we had close to $10 million in timber cutting bonds, plus about $1.1 million in self-insurance bond. Upon filing bankruptcy, these contracts came into jeopardy. Most of the timber cutting bonds and performance guarantee bonds began to look unstable.

The log and lumber market deteriorated so far down that National Forest timber offered at $15 to $20/M received no bids.

Tom Mayr, with the help of John Strasburger of Short & Cressman, our attorneys, searched out workable contract cancellations on many contracts. Fortunately we had most of the roads needed to log the State timber sales completed and paid for and, in effect, we traded our timber deposits and completed roads for release of some of the timber sales. Also, we bought out of some sales under a law designed for that purpose, and we bought out of two Forest Service sales at a cost of improvements plus $10/M under a Federal law.

On the two timber sales located on the Makah Reservation, we had a very difficult time. Had the bond been $25,000 or $35,000 they would have been easier to deal with, but they were bound and determined to get the $400,000 cash bond. Tom finally engaged Saltman & Stevens for us and they did save us a little more than their attorney fee. Later I met Saltman and I was not too impressed with him, a typical Washington, D.C. attorney engaged in legal work for private clients.

By the middle of 1987 we were settled up with Federal Land Bank, State of Washington Department of Natural Resources, the U.S. Forest

Service, the Bureau of Indian Affairs for the Makah tribe, and the State Department of Labor & Industries. It would take both of the remaining properties we had pledged to United Pacific Bonding Co. to pay off the losses. Some of the losses were written into the plan.

Rainier Bank wanted written into the plan a schedule which would pay them and their stockholders 100 percent plus interest. Again, we had no money to fight with so we gave way. The rest of the creditors agreed to the same terms and conditions.

Before we could reorganize, the bank wanted me out as operating executive of the company. When a person is down, it is pretty easy to kick them out, which was pretty well done to me by the end of 1987, but I couldn't work at a place where I was ignored and continuously working outside of the team.

In the winter of 1985-'86 we came within an inch of going under completely. We had no logs and no lumber inventory control at all, just lots of lap-top paper work inherited from M.T.A.

In sawmill management, there are really only a few things that affect your financial position: Inventory of logs and lumber; and log cost, milling cost and the selling price of lumber and pulp chips.

Years before, when our company was a stockholder in Blagen Mill Co., Mr. Len Skalley was in charge of the mill. He submitted a two-page, typed statement of the operation which clearly showed condition of the company and results of the preceding month. I thought that was enough.

I think the bank wanted the huge M.T.A. reports because they love lots of paper.

After finishing logging of the Elastic timber sale, MB Logging started falling and bucking on the Cougar timber sale, which we had purchased in 1977. On the timber sale there was about 5 million board feet of old-growth Douglas fir with some cedar and about 15 million board feet of average hemlock remaining. We were able to get this timber on the new plan, which was part of the new Federal recovery plan passed through Congress, allowing a five-year extension on the contract without penalty. Logging was carried out on one side using the 208 Washington.

Tom was able to sell the old-growth Douglas fir to Avison Mill Co. near Portland at a very good price. The remaining timber on this sale was a Godsend. It made enough money to set up the logging company.

It also got the company back into purchasing of timber sales. The first two were "Backstretch" and "Matheny Breaks." Inevitably the lumber prices will firm up again and probably go even higher than they were in the 1978-'79 period, but purchase price on these two sales was

The rigging crew. Photo © John Tylczak

approximately $50/M, $180/M lower than the price paid in 1980 before the lumber market crashed.

In the post-bankruptcy years, the pressure from Rainier Bank began to mellow a bit, especially after our final settlement with the Federal Land Bank and the Bureau of Indian Affairs. Rainier Bank surrendered their second position on the huge Federal Land Bank timber loan.

Logging at the Cougar timber sale did the most during 1986-'87 to increase cash flow and profit. That allowed MB Logging to hold back invoicing the parent company on log sales, thereby increasing cash flow. Continued strengthening of the market allowed MB Logging to make profit on both Backstretch and Matheny Breaks timber sales.

The thousands of acres of timber land we worked so hard to put together during the years from 1944 through 1980 have now all been sold by the Federal Land Bank to various new owners, including John Hancock Insurance Co., Equitable Life Insurance Co., Quinault Indian Nation, and Golden Springs Co. The Vesta Creek cottonwood plantation was purchased by Ray Oatfield.

Much of the land has already been logged again and there is nothing left but brush. Areas that were logged in the 1930s and 1940s have all been logged over again. All the properties which our company owned and had not logged have now been logged.

Teaching my children the art of planting trees and tree farming will be of no use to them now. Maybe they will have some pleasant memories of the days when they were children and we used to go up to the tree farm on Sundays and picnic on the sand bar above Tree Farm 5, the bridge, and the treehouse we built along the old railroad grade in Section 13.

After the purchase of Twin Harbor's mill and timber assets, we had set up a new corporation, Willapa Timber Co., in which 50 per cent of the common stock was held by our children and 50 per cent by Marzell's children. Marzell, Jennie and I owned preferred shares which paid interest for only a short while until our company filed bankruptcy. After bankruptcy the stock became worthless and this affected adversely the family relationship with me.

Looking back in retrospect on all those years, it is hard to understand how Marzell and I worked together through the trial and tribulation of being in the "left-over" end of the logging business and the mill business, and then, after fifty years, make such poor decisions one after the other. So now, after so many years, I was really losing interest in the company and was ready to accept my papers. I felt as droopy as the lowly hemlock tree, the Cinderella tree.

Chapter 9: Into the Nineties

Moving On

JENNIE AND I ARE IN OUR FIFTIETH YEAR of marriage. Our children are gone from home and all are living their own lives. We have thirteen grandchildren.

Catherine Susan and her husband Oliver have Chris Edwin, a student at San Jose State in California and Todd Allen, enlisted in the U.S. Navy and stationed on board a submarine tender stationed in Guam. Cathy works as receptionist in Mayr Brothers main office. Ole and his sister Vicki operate Port Machine Works very successfully.

Patricia Ann lives in Beaverton, Oregon, and works in the personnel department of Western Medical Services, which is the firm responsible for furnishing personnel for the medical profession of the greater Portland area hospitals. Ron Webb works as head saw filer in the Mayr Brothers Logging Co. saw mill. They have three children, Derek Andrew, age 20, Miranda Maureen, age 13, and Silas Mayr Webb, age ten. Derek has an EM-CADD draftsman degree and is advancing his education.

Michael works at Mayr Brothers Logging Co. log yard marking log sorts, tagging for recording and identification. His wife Holly works for the Elma Post Office and has a rural delivery route. They have two boys: Jeremy Michael, age 11; and John MacKenzie, age five. They have a horse ranch just above the school in the Wishkah Valley.

Daniel and Cindy have Joshua Joseph, age 11 and a great little leaguer; Ami Michelle, age nine, is a great swimmer; Lindsey Lee, age seven and ready to begin school; and Chanyn Cheree, age five. They are all great children. Daniel is in lumber procurement and sales for Pooser Lumber Company of Sacramento, California.

David and Theresa were married last August and live on a place David built up, located in Cosmopolis, Washington. David has an automotive performance shop, located just off Curtis Boulevard in South Aberdeen, where he does high quality automobile work. Theresa works in lumber sales for Mayr Brother's Logging Co. in the Lumber Department at North Hoquiam.

Caroline Marie lives in Tacoma and works at Jet Equipment in the sales department. We frequently have lunch with her and Suzanne at the Sheraton when we go for Jennie's appointments with Dr. C. Caggiano, D.P.M. in Tacoma.

Suzanne Theresa lives in Fircrest, Washington, near Tacoma, and works at Multicare Medical Group-Fund Development, which is the fund-raising group for Mary Bridge Children's Hospital and Tacoma General. Suzanne is married to Lt. David Lee, U.S. Army Infantry. They have two sons; Patrick John and Steven Werner.

John Richard, our number eight child, lives in San Francisco, California, and belongs to the DeLancy Street Foundation group. This group has been very successful with drug and alcohol addiction in young people. DeLancy Street Foundation receives no government financial help, nor does it belong to any sectarian group. As their saying is, "We make it on our own." Jennie and I, together with John's nephew and nieces, have visited John and the foundation many times.

One of the biggest ventures of the DeLancy Street Foundation is the annual Christmas tree sales every year. They have a near-monopoly on Christmas tree sales in the Bay area. John worked with this group last Christmas and hopes to be part of the program this year. They pre-order trees for the large bank lobbies and the hotels, ordered to fit each customer. Orders must be in by September 30. All Christmas trees furnished by DeLancy Street Foundation are purchased from Christmas tree growers located in western Oregon.

After Christmas, John will enroll in the City of San Francisco College and take two courses. John is confident he will be able to get a good job in marketing. DeLancy people are very sought after in California; the employer feels he is getting the best.

Marzell and Snevah's children both are college graduates and have degrees: Tom in Forestry and Management; Mary in Home Economics and Business. Both work in Mayr Brothers' Logging Co., Inc., management.

Tom married Theresa Stebbins, and they live on the old home place on the Wishkah. The have Rachell Annette, age 13; Anthony Marzell, age ten; and Jamie Marie, age six.

Mary married Steve Nelson, who attended the University of Idaho at Moscow, Idaho. Steve also works in management of the company in the chip mill and log transportation. Steve and Mary have Amy Margaret, age 11; Heather Elizabeth, age nine; Alyssa May, age seven; and Parker Steven, age five. All are beautiful children.

Francis Hannick married Carol Eliaason from Hoquiam. Later Francis became a Mathematics Professor in Minnesota University at Mankato, Minnesota, and Carol is a librarian in the same city.

Steven Hannick married Nancy Israel. They live in a suburb of Chicago. Steven works for a large pharmaceutical company defining and developing new medical drugs. Nancy and Steven have Jessica Helen, age seven; and Michael Joseph, age four.

Barbara Hannick is married to George T. Reich. They live in Bellevue, Washington. George works for Bonneville Power Administration. They have Heidi Jean, age seven; and Maxwell T., age three-and-a-half years.

The families have held together well, in spite of the divergent background in the fourth and fifth generation, and especially considering the dilemma we found ourselves in at bankruptcy.

Most manufacturing companies early in their beginning get into some kind of problem. Rarely do they go for more than 15 to 20 years before some part of planning does not work out. The longer a business goes before getting into serious problems, the harder it is to correct and make the necessary changes. Mayr Bros. Logging Co. was the last logging and sawmilling company successfully built from the horse logging to the top, and we created a very substantial payroll for the twin harbors.

In my estimation, our greatest loss was the 45,000-acre tree farm. Our tree farm was one of the finest in the industry, with 350 million board feet of marketable timber, thousands of acres of plantations, hundreds of miles of graveled roads. It was the end of an era. This land could never be replaced.

When Marzell, Jennie and I started building our tree farm, there were hundreds of thousands of acres of the very best timber land available. Our timber land base began near where we grew up, and that land on the Wishkah was some of the best. We learned to cruise and evaluate timber and land, how to locate property lines, and I got great satisfaction following growth patterns as the timber properties developed, and after 40 years the trees were harvested and the land was replanted.

After bankruptcy, when we deeded the land to the Federal Land Bank of Spokane, we also signed off our right of one year to redeem the property. The bank used a strangle-hold by threat of taking our homes to force the agreement, and I believe there was collusion between the Federal Land Bank and Rainier Bank against us. Within a very short period of time, the Land Bank set up an office in Hoquiam and began selling off the properties at the going market, which was horribly low, I think less than one-third of its value two years later, though we have no knowledge as to the prices the bank liquidated our timber property at.

After our families went through the bitter three-year struggle to pull the company out of bankruptcy and save what the lenders would allow us to keep, breathing became a little easier and it became easier to sleep until 5 or 6 a.m. Little by little the company made progress. The Cinderella tree found its place in the market, and there was hope that we could regain our place in the industry and in the community of Grays Harbor.

We must move forward, no matter what our families went through. It is time to turn pages and let the past be gone, time to look at the future. Young people have new ideas, and face new pitfalls. What career or new business is there to go into, when, in the very middle of the timber country, sawmills are barely competitive for only a small part of the timber resources? When we allowed ourselves to become a tree farm for Asia, I think we made a mistake. Exporting raw logs, though we shipped millions of board feet of the same, might better be banned.

We suffered at the loss of our privately owned timber supply and shrinking supply of public timber caused by environmental issues and other national issues which look at tree harvesting as a destructive act which harms the environment. It is time to give up the timber business or build a super efficient log handling, lumber manufacturing factory.

We were always elated to see the thousands of acres of young, thrifty forest growing on the acres we logged in the old-growth forests north of Hoquiam. The young forests are resplendent in their glittering green foliage and are home for deer, Roosevelt elk, cougar, bobcats, and lots of black bear. The streams that always had fish still have fish. Logging and converting the land to a new, rotating forest has not and will not harm wildlife and fisheries, as harvest is carried on by good forest management directed by the U.S. Forestry Service and Washington State Department of Natural Resources.

Our Creator put all kinds of resources on this Earth for man, His creation, to use and develop the world from pole to pole. The water and heat and light from the sun makes things grow. The world constantly changes. What was here today is gone tomorrow. A constant rotation of crops of all kinds, including timber trees, improves the environment for man and wildlife.

As the North American continent developed and civilization covered from ocean to ocean and from the tropics to the Arctic Circle, sacrifices were made. The grizzly bear, the plains buffalo, the grey timber wolf, the carrier pigeon, are some of the species that have been "hauled in" or no longer exist in the wild. Grey wolves are still found in the cattle country of Cariboo of British Columbia, Canada. Grizzly bears are still found in parts of Montana and Idaho.

It is my observation, which is the result of many years in the forests of western Washington, that forest management must go on and expand on all lands, both public and private, even in the National Parks, to recover the waste of resource.

What brings me to this conclusion is the total environmental soundness of growing commercial forest trees and the environmental soundness of the products produced from trees and their uses.

A growing tree holds water and soil together during extremes of weather. Trees give shade, and produce food for birds and animals. Timber growth adds to the oxygen in the air, produced by the thrifty growth of healthy young trees, which consume carbon dioxide to a far greater extent than old, decadent stands of trees.

In their best environment, trees are harvested and continue giving good use for fuel, building material, myriad wood pulp products and paper, fibers for cloth. Wood products, especially some of the highly sensitive ones like toilet paper, are pretty difficult to replace with any other natural or man-made fibers.

The use of wood will continue to expand. Wood feels warm to the touch, is versatile, and, properly applied, can last hundreds of years. It can be reused time and again and finally makes a good fuel.

When wood was plentiful and the price was low, there was no thought of pressure treating to protect wood from fungus and from insects of many kinds. Now the paper mills are seeking ways to recycle used paper fibers to be used in manufacture of newsprint and packaging crates like egg cartons.

The facts are indeed a far cry from the outlook and direction given by the anti-timber harvest people. Sometimes, I wonder if they have ever studied out the life and times of our northwest forest species and forest sites, and if they have, what alternate materials they propose be substituted for wood and wood fiber? Wood as a construction material has been used by man from the oldest times. As civilization progressed, better uses and ways were discovered.

Production of wood by wise use and cultivation of growing timber crops improves the living conditions on our planet. The timber growing capacity of the climate and the soil are factors that, properly cared for, will produce timber crops equal to or better than the original wild, old growth stands available before the early settlers came to develop the land.

The only premise left for a common sense person speaking against management and harvest of old-growth, decadent timber is that the huge old-growth tree is a marvel to look at and admire. There are already millions of acres set aside for sight-seeing located in the National and State Parks of Washington and Oregon.

In our area can be found timber stands of Douglas fir, hemlock and spruce 100-plus years of age which are up to 36 inches, diameter breast height and four logs in height, scaling up to five thousand board feet per tree. These second-growth trees, between planting and 50 years of age, are establishing a root system and a stem and their place in the forest floor for future growth. The greatest annual volume increase comes between 50 and 80 years. After 80 years, the western conifer begins to slow and the growth rings narrow down and the tree clears up from the early year branches.

The optimum period to harvest this type of soft wood species is at 80 years, not 150-200 years or older. Harvesting trees younger than 45 years is a sheer waste of the growing capacity of the soil and will cause an extreme shortage of available timber because harvest at 45 years yields only 35 percent of usable wood compared to harvest at 80 years, an age that also produces large saw logs suitable for most lumber uses, including structural timbers and glue-lam material.

All of the old-growth stands of timber need to be harvested and the land replanted in order to increase the timber growing land base and eliminate the need to harvest immature second-growth. The increased land base by combining additional acreage now under spotted owl and other "goody's" management and the increase in yield from the extra growth on timber growing to 80 years will put the Pacific Northwest far in the lead for timber resources.

In fifty years the old forest will be gone whether it is logged or not. It is so far advanced into decadence that only brush and sub-species will survive. The old forest will need very heavy silvicultural applications to get the land back into production of usable timber species, and only fire produces the environment to properly begin a new crop of Douglas fir.

The area we live and work in is known as the temperate zone conifer forest. This area extends from the Pacific Ocean to the Cascade Mountains up to elevation of 4,500 feet. The entire area from the

Canadian border to the Columbia River was at one time covered with dense conifer forest. Out on the slopes near the Pacific Ocean there is now evidence that it is the best site for Douglas fir. It could never get started there until the ground was cleared up of brush by fire.

There were some areas partly burned over by the native people to create hunting areas for game and berry fields. Heavy, dense forest areas do not create food for man, animal or bird because the lack of sunlight causes a lack of protein which directly or indirectly affects all creatures living under this forest canopy.

True, there are marginal creatures found under this dense, wet, cold forest canopy, but let sunlight enter and you see food plants thrive and create protein energy for the chain of life. The Olympic Peninsula always was a poor place for birds to even survive. That is why there are so few spotted owls in the west slope of the Olympics. Even the black-tailed deer found is scrawny compared to the mule deer found on the eastern Oregon slope where there is sunlight.

To get the timber off the stump and down to the sawmill is a large part of the overall production in manufacture of lumber, plywood and other products. Besides timber and equipment, it takes people to make an operation functional. In the forest work, it takes people that like the outdoors and don't mind the long rainy season, which we have plenty of out here on this north Washington coast.

The loggers working in these forests get so they like the weather, including the winter's snow and mud, and the summer's dust. Then comes fire season and "hoot-owl," which means leaving home around 2:30 a.m., hitting the woods at daylight and getting out at 1 p.m. before the heat of the day makes for critical forest fire danger.

Loggers work on some pretty clammy days and then comes a day the sky is clear and blue and you'd swear, if you were high enough, you could see clear to China, and the logger says, "I would not trade one beautiful day like this for the whole world."

All through the year when we were young and worked right alongside the men, we wore the heavy leather shoes with high sides (ten inches) and soles with steel spikes, called "caulk boots," or, in logger lingo "cork boots," although there was nothing light about them. On rainy days, we wore heavy denim pants that were waterproof, more or less (mostly less), and called them "tin pants" because they were so stiff they would stand up by themselves. In winter, there were damn heavy black wool long johns, and they kept you warm on rainy, sleety days.

The truck drivers, for the most part, had it the best. They could get dried out driving to the log dump, although there can be a lot of stress

Marzell and I together in the yard at Hoquiam.

driving a big six-wheeler pulling a four-wheel trailer loaded with 60,000 pounds of logs and making it back to the woods loading area before quitting time, all the while on the alert while going down the highway, meeting people coming in cars who have no idea of what was going on.

After many years of selling logs we became acquainted with the chip mill and sawmill business. That is a different ball game. Lloyd Allen said one day after we had been in the saw mill business for awhile, "Werner, how do you like the sawmill business compared to logging?"

I said, "Lloyd, I guess it's okay."

Then said Lloyd, "Well, at least in the sawmill business you have everything right in one place, including tools and equipment."

The sawmill business is close to town and this made it fairly easy to get parts or mechanical help quickly. Sawmill people, I found to be just as fine a class of people as you might find anywhere, honest, and hard working. In the sawmill and chip mill, the people would wear hard-toe shoes to protect their feet, and, of course, the ever present hard hat.

About half the crew eat lunch at a fast-food place rather than carry a "mamma-made lunch," but loggers out in the foothills always carry a lunch pail and at lunch time, they will sit on a log, stump or the wet ground and munch their lunch. Then things look better because they are on the downhill side of the day.

In the early days in the mills, the setter rode the carriage back and forth and made the sets as per the sawyer's instruction, and in those days, I spent many an hour in our log customer's sawmills, and as a result I knew a lot about the sawmill business long before we started building our mill. Those big old mills used a 10-foot band wheel-size head rig with 8-foot back-stand space, and could cut 8-foot diameter logs and make 36-by-36-inch timbers 40 feet long.

Like all things in life, this comes to the end. It's time to wind down. Whatever errors I have made, wrong dates, wrong people or whatever, time will erase them. Who will be left to remember?

So many things in life often are beyond our control. We hear so much these days of a career, and so many things are more important and often go completely unnoticed.

For instance, we always had a milk cow and maybe 50 to 100 chickens, and Mom sold eggs so that gave us more work that had to be done summer or winter.

Dad loved hard cider and made a big barrel of it every fall.

We would all get together at the folk's place on Sundays and Christmas.

At least, like the Cinderella tree in the second-growth forest, we prevailed.

The End

A Family Tree

Mayr Family

Father   **Mother**

Mathilda Sommer
Born 1877. Died 1963

Father side		Mother side
Werner Sommer *Born 1845. Died 1921*		H.A. Baucher *Born 1851. Died 1884*
Joseph Anton Sommer *Born 1817*		Louise Alliegor *Born 1814*
Anton Sommer *Born 1789*	*Married 1814*	B.A. Reich *Born 1795*
Joseph Sommer *Born 1716*		

Marzellinous Mayr
Born 1877. Died 1952

Kajetan Mayr *Born 1849. Died 1920*	Walburga Krumm *Born 1845. Died 1920*
Alexander Mayr	Crezenzia

*Marzellinous and Mathilda married at
Maria Insedlen, Switzerland, in 1909.*

Francis Louis Hannick
Born 1909. Died 1981

Joseph M. Hannick *B. Sept. 28, 1879. D. Jan. 14, 1965*	Adeline Hannick *Born 1889. Died 1965*
John L. Ditgen *Born 1863*	Anna M. Mecters Sheimer *Born 1854. Died 1938*

*Francis L. Hannick and Hedwig M. Mayr
were married June, 1941.*

House Family

Father

Mother

Richard House
(Husband of Margaret Mayr)
Born 1901. Died 1963

Francis Dionne
Born 1922. Died 1952

Adolphus Chartes Dionne
*Born 1887 at Red Rapids, Isladwin,
New Brunswick, Canada*

Catherine Mary Ryan
*Born 1888 at
Johnville, New Brunswick*

Snevah Havens
Born 1913

Jesse L. Havens
*Born 1883. Died 1949 at
Colby, Michigan*

Mary Theresa Smith
Born 1883. Died 1953

*Francis M. Dionne and Marzell T. Mayr were married 1949.
Snevah Havens and Marzell T. Mayr married Sept. 9, 1954.*

Jennie K. Brolin
Born 1922

Helmer S. Brolin
Born 1890. Died 1935

Gustav Brolin
(Father of Helmer S. Brolin)
Born 1860

Brita Strindberg
Born 1889. Died 1958

Emma Strindberg
Born 1855. Died 1947

Brita Strindberg and Helmer Brolin were married 1919.

Names of 8th Generation from Joseph Sommer

Margaret Mayr married **Richard House** 1950. No children.

Werner Mayr married **Jennie K. Brolin** 1941. Children
  Catherine
  Patricia
  Michael
  Daniel
  David
  Caroline
  Suzanne
  John

Marzell Mayr married **Francis Dionne** 1949. Children
  Mary
  Thomas
Francis Dionne died 1952.

Marzell Mayr married **Snevah Havens** 1954. No children.

Hedwig Mayr married **Francis L. Hannick** 1941. Children:
  Francis
  Steven
  Barbara

To Order Additional Copies of
THE CINDERELLA TREE
Use This Handy Coupon

Please send me _____ copy/copies of *The Cinderella Tree* at $14.95 per copy. (To defray shipping, please add $2 for the first book ordered and $1 for each additional book.) Send check or money order — no cash or C.O.D.'s, please.

Name_____

Street_____

City_____**State**_____**ZIP**_____

Mail To:
Keokee Co. Publishing
P.O. Box 722
Sandpoint, ID 83864
Please allow three weeks for delivery.

•

Please mark the boxes below to indicate books of interest to you:

❑ Northwest history, nature and natural history books.
❑ Northwest travel and outdoor guidebooks.
❑ Poetry and fiction by Northwest authors.

•

Also available from Keokee Co. Publishing:

The Klockmann Diary by A.K. Klockmann,
Edited By Chris Bessler.
An exciting personal history of mining and old times in turn-of-the-century Idaho. Nonfiction.
ISBN 1-879628-00-7/ $9.95.

The Blascomb Family Chronicles by M.R. Compton, Jr.
Tales of Montana, and ordinary people leading extraordinary lives, well-told by a native Northwest storyteller. Fiction.
Caleb's Miracle - $4.95
Coming soon: Jason's Passage